# WE DIG AMMONITES

# WE DIG AMMONITES

## Fact, Folklore, and History

## JODI SUMMERS

Forewords by Andy Secher and Neil H. Landman

Columbia University Press
*Publishers Since 1893*
New York    Chichester, West Sussex
cup.columbia.edu

Library of Congress Cataloging-in-Publication Data
Names: Summers, Jodi author
Title: We dig ammonites : fact, folklore, and history / Jodi Summers.
Description: New York : Columbia University Press, [2025] | Includes bibliographical references and index.
Identifiers: LCCN 2025004027 | ISBN 9780231213929 hardback | ISBN 9780231560108 ebook
Subjects: LCSH: Ammonoidea
Classification: LCC QE807.A5 S887 2025
LC record available at https://lccn.loc.gov/2025004027


Printed in the United States of America


Cover photo: Jodi Summers


GPSR Authorized Representative: Easy Access System Europe, Mustamäe tee 50,
10621 Tallinn, Estonia, gpsr.requests@easproject.com

**COVER PAGE IMAGE**: An Early Jurassic spiral ammonite, this species of *Lytoceras* had Saturn-like concentric rings hugging the wide outer coil.
**Lytoceras fimbriatum (Sowerby, 1817)**
Orders: Ammonitida
**Passaloteuthis Lissajous, 1915**
Order: Belemnitida.
Appearance and lifespan: Early Jurassic, 189.6 to 183 MYA. Location: West Dorset, England, United Kingdom. Size: The ammonite is 15 cm.

**TITLE PAGE IMAGE**: A cluster of iridescent *Cleoniceras* ammonites.
**Cleoniceras Parona & Bonarelli, 1897**
Order: Ammonitida. Appearance and lifespan: Late Cretaceous, 105 to 100 MYA. Location: Mahajanga, Madagascar. Size: largest is 12 cm.

This book has been created for the enjoyment of

ammonoid fans everywhere.

Thank you to my family for their unyielding enthusiasm

and encouragement.

With love and appreciation to Andy Secher; without your

inspiration and support, this project would probably have

had a very different tone. I am eternally grateful that you

introduced me to the pleasures of paleontology.

And a special thank you to my editors,

Dr. Christian Klug and Neal L. Larson. I believe that

together we have created a worthy salute to this

venerable cephalopod.

*Speetoniceras* split, with the shell exterior encased in pyrite.

**Speetoniceras Spath, 1924**
Order: Ammonitida. Appearance and lifespan: Early Cretaceous, 132.9 to 129.4 MYA.
Location: Volga River, Ulyanovsk Region, Russia. Size: 20 cm.

# Contents

***Deshayesites deshayesi*** (**Leymerie, 1840**) ammonite spray.
Order: Ammonitida. Appearance and lifespan: Early Cretaceous, 125–113 MYA. Location: Sartow, Russia. Size: largest is 4 cm.

**Hoplites dentatus**
Order: Ammonitida. Appearance and lifespan: Early Cretaceous, 113 to 100 MYA. Location: Courcelles, France. Size: 12 cm.

# Preface

As one the most prolific fossils in the geologic record, *Ammonoidea* has been discovered in sedimentary strata from Afghanistan to Zimbabwe. Their smart spiral shells have inspired myths, legends, gods, saints, and science. They are collected, admired, appreciated, and displayed throughout the world for personal pleasure and cultural beliefs. The stories of their paleontological importance and social relevance deserve to be preserved for history. Indeed, these cephalopods are amazing creatures, and within these pages we'll explore why this is true.

With a passing glance, these ancient sea animals remind us of things we see daily. Stone snail shells with an inverted whorl, we notice the ammonite's spiral shape everywhere—from rolls of paper towels to corkscrews, helixes, and hurricanes. That's why the familiar, curled conchs of ammonoids make them one of the most recognizable and pleasing of all fossil forms. But beyond being lovely to look at, these unique creatures serve a purpose. During their nearly 400 million years on our planet, *Ammonoidea* produced more than 10,000 genera and an infinitely greater number of species. Their shell details evolved so quickly—and fossilized so well—that geologists can now identify the age of layers of rock according to the types and combinations of ammonoids found in that stone strata.

This hopefully enjoyable general science book, geared expressedly for the mainstream market, shares the legacy of *Ammonoidea* in words and pictures. Hello world! The goal of *We Dig Ammonites* is to be an outlet to better appreciate these magnificent and telling prehistoric timepieces.

Had I grown up in a fossil-rich area, this story would likely be different. Raised on the glacial moraine of Long Island, my love of ammonites began with seashells. Ever since I discovered those magnificent little treasures that washed up on the seashore, I was fascinated by their colors and variety. But while I adored the conch, I was none too enthusiastic about the slimy creatures that had previously lived inside and whose remains could quickly stink up a room. My prized shells—some found, and some purchased—were carefully stored in a small hardware utility cabinet with plastic drawers emblazoned with labels that indicated the fauna's common name and locale. I brought some of these conchs into second grade show-and-tell and my classmates oohed and ahh-hed over the instantly alluring heart cockles and astutely noted that chocolate humpback cowries were Bambi-spotted.

Always a prolific creator, particularly with a camera, I photographed and wrote elementary school reports on things like carrier snails, why scallops are so brightly colored, and the unique counterclockwise development of the lightning whelk. That old cliché, "A picture is worth a thousand words," is so true. Those 8 × 10 Fotomat prints of shells on vibrant green grass or on a black tar driveway always got me an A. Sometimes my teachers didn't even comment on the text, making me wonder if they'd even bothered to read the composition. My collection kept me busy and out of trouble, so everyone encouraged my obsession. My grandparents would take my cousin and me to the Brooklyn Aquarium, where the gift shop featured an amazing variety of shells, and we'd often get to choose a souvenir to take home. I can remember being absolutely delighted over a spider conch or a green turban

shell; ahh, the simple pleasures of youth! During my teenage years, on the shelf next to my music collection, were such curiously named ocean collectables as queen conchs, flamingo tongues, and carrier snails. They were proudly on display accompanied by spectacular photo books on the subject of shells. That collection sparked grand ideas about how to share the beauty of nature in two dimensions.

Thanks to a chance meeting with the venerable Andy Secher during my impressionable college years, I was introduced to the fascinating field of paleontology. He collects trilobites, those buggy-looking extinct marine arthropods that men seem to find so appealing, while women tend to think they look like cockroaches. Because he is such a brilliant guy, I often take Andy's imparted wisdom to heart. He believed that interesting people collected interesting things. He familiarized me with his favorite Manhattan fossil haunts and we began attending regional rock and mineral shows. There, I found my niche. My focus shifted from seashells to a related but somewhat more prosaic interest: *Ammonoidea*. I appreciated that their fossilized forms never had living creatures inside their conchs. Whether it was the snake-headed *Hildoceras* of legend and fable, the simplicity of a primitive goniatite, or a *Perisphinctes* with preserved soft-body parts, ammonoids and their stories appealed to me. There was something about the myriad of Paleozoic and Mesozoic variations and their elegant forms that charmed me and my photographer's perspective. As soon as I developed an interest in these cephalopods, I realized I wasn't alone. There are collectors all over the world. Acquisitors like me appreciate the benefits of extraordinary variety. Others find them an aesthetically pleasing reminder of primaeval times. Some enthusiasts are drawn by certain strange pathologies, biostratigraphy, or other arcane academic insights the specimens may reveal. Perhaps most of all, a great number of aficionados have realized that searching for ammonites is just plain fun.

Those who dig take great pride and pleasure going out into nature and uncovering something that has never before been seen by human eyes. The more style-conscious appreciate these cephalopods as an attractive piece of decorative art that adds a sophisticated touch to their home. Whatever may be the reason for liking them, ammonoids were extraordinary sea creatures. Known to most of the modern world as ammonites, the scientific term is actually *Ammonoidea*. Simpler forms known as ammonoids emerged during the Paleozoic Era. The more evolved ammonite order emerged in the Mesozoic. Technically speaking, *Ammonoidea* and "ammonoid" are all-encompassing terms. The popular label "ammonite" actually refers only to the conched Mesozoic varieties.

Back in those halcyon Cambrian times, ammonoids were barely larger than millimeter microfossils—just tiny creatures with straightish, sometimes softish, uncoiled shells. These prolific predatory molluscs were a squid-like, soft-bodied animal that was shielded by a firm outer shell. Though there is not enough information to know for sure, the nonbiomineralized creature is believed to have had a relatively consistent morphology, while its protective conch continually evolved. A supremely successful form of sea life with no apparent predators, ammonoids grew in size and number, rapidly radiating to fill countless benthic biozones. By the beginning of the Mesozoic Era, these voracious cephalopods dominated the seas, and there was plenty of ocean to dominate. The supercontinent known as Pangea was forming, and it covered about a third of the planet's surface area. This left substantial sections of today's landmasses to serve as underwater breeding grounds for burgeoning forms of sea life. Suddenly, the waters were filled with coiled cephalopods of many shapes, styles, and dietary habits. During their oceanic reign, ammonoids enjoyed a varied regimen of smaller creatures, which allowed them to thrive in a plethora of ecological niches. Some of the more ambitious genera, most notably species of the planispiral *Parapuzosia* and the trombone-shaped *Diplomoceras*, had shells that grew up to two meters (six feet) in size. Think of them as being akin to giant squids with an exterior conch.

Today, ammonoids are mementoes of an ancient aquatic past. Shells of various styles and ages are preserved on all seven continents. Thanks to plate tectonics, marine life like ammonoids, crinoids, and trilobites may now be found buried in mountain ranges or in locations hundreds of kilometers from the nearest saltwater sea. Indeed, these coiled treasures have been uncovered just about everywhere that sedimentary outcrops of the right age exist. Devonian and Carboniferous ammonoids are seen throughout Morocco's daunting Atlas Mountains. Cretaceous ammonites have emerged from Peru's cordillera, and magical pockets abound in the North American Rockies. Final remnants of the subclass have been discovered on the desolate islands near Antarctica. Specimens of all ages are found on hills and in valleys throughout Europe, and they weather out of the shoreside cliffs of merry old England on a daily basis. Depending upon where you live, *Ammonoidea* can even be uncovered in some fortunate enthusiast's backyards. If there are Mesozoic limestone escarpments about, it's likely that some species of these prolific cephalopods are lurking within those preservations.

In this book, we will get down and dirty in key prospecting locations and stay clean and pristine as we travel to UNESCO World Heritage Sites. Near the world's most outstanding ammonite preservations, are locations where the layman can legally unearth exceptional cephalopods.

We will explore and collect from the cephalopod-rich mountains of Pakistan, Ecuador,

The author with an example of the largest genus of ammonite.

***Parapuzosia*** **(Nowak 1913)** Order: Ammonitida. Appearance and lifespan: Late Cretaceous, 83.5 to 70.6 MYA. Location: Maverick County, Texas, United States. Size: approximately 150 cm.

Image credit: Andy Secher

Morocco, and Tibet. We'll put on hard hats and rubber boots prior to stepping into mines and mud bogs throughout Poland, Germany, and Japan. We'll traverse the karst structures of China and Spain, and search for fossils along the coasts of England, France, and Portugal. In these pages we will discover lost watercourses like the Western Interior Seaway of North America, and go beyond the Cretaceous extinction event in Denmark, Antarctica, and even exotic New Jersey.

Fascinating as these cephalopods and their storied sojourn may be, I never seriously considered taking the arduous scholarly discourse needed to become a curator at one of the leading American museums. Rather, my interest is strictly from an enthusiastic acquisitor's perspective. For that reason, when required to identify a specimen (an imperfect art at best), I got as deep as the genus. I leave it to the academicians to debate, name, and rename species. With

degrees and awards in journalism and photography from the acclaimed Walter Cronkite School of Journalism at Arizona State University, I've always felt I'd had the ability to master whatever might interest me—and, in all modesty, I have. After decades of obsessing over these fascinating fossilized creatures, my knowledge and collecting ability is at such an elevated level that I am now proudly a field associate in paleontology at the American Museum of Natural History in New York City.

When I started out on this excursion with a perfect polished Moroccan goniatite, little did I realize that it would grow into an interest that I would be able to enjoy throughout my entire adult life. As I amassed both knowledge and specimens, the question then became, with so many types of *Ammonoidea*, something new: Was focusing just on this cephalopod enough specialization to become an intelligent collector? Or should I micromanage a subclass with so

When the exterior nacre of the shell is buffed off, the ammonoid's chamber construction is better revealed.

**Gonioclymenia Hyatt, 1884**
Order: Clymeniida. Appearance and lifespan: Late Devonian 364.7 to 360.7 MYA. Location: Errachidia Province, Drâa-Tafilalet Region, Morocco. Size: 24 cm.

An ammonite fossil in situ along the Lyme Regis coastline.

Image credit: FluffyBiscuit, CC BY 3.0, via Wikimedia Commons.

many specimens by concentrating on period, location, or family instead? As I have never lived in an area where examples are easily found, I decided to assume a more expansive perspective. I assembled a coterie of ammonoid fossils that appealed to my photographer's eye because of their extraordinary preservations. Essentially, *We Dig Ammonites* is as much a photo book as it is a history of ammonites and exceptional places to find them.

As an amateur digger who seldom uncovers the quality of specimens I have in my collection, I have acquired many of these primal treasures at the planet's major fossil shows. Worldwide events, paleontology fairs annually occur in locales such as Tucson, Denver, Tokyo, Munich, and even in the charming village of Sainte-Marie-aux-Mines in the Alsace region of France. Certainly, the opportunity to travel to interesting places in pursuit of *Ammonoidea* adds another layer of excitement to my ongoing quest to compile one of the world's preeminent collections. I am pleased to note that if you don't see a photo credit, that specimen has likely been drawn from my own extensive collection, and its image captured by me.

Throughout this book we shall adventure with a formidable cast of fossil celebrities, specialists, academicians, museum curators, collectors, vendors, and scientists. Together we will journey into the field to find environments featuring ammonites in situ. And we shall visit the hallowed halls of public and private institutions which feature *Ammonoidea* collections.

*We Dig Ammonites* is equally inspired by these amazing cephalopods, my love of travel, and the wondrous seashell books of my childhood. My intent is a compendium that captures the allure of these significant specimens, highlights their relevance to the world, and shares some of the most fascinating places to experience and discover them.

Enjoy the journey!

*Perisphinctes* **Waagen, 1869**
Order: Ammonitida. Appearance and lifespan: Middle Jurassic, 163.5 to 157.3 MYA. Location: Tulear, Madagascar. Size: 8 cm.

# Foreword

Andy Secher

've known Jodi Summers since before she liked ammonites. Back then her primary focus of attention seemed to be evenly divided between digesting any new music released by Elton John and deciding what distant ports of call she could visit on a seemingly limited budget. I certainly never imagined that she would soon evolve to share my passion for the natural sciences—fossils, in particular. Yet slowly but surely the wonders of paleontology began to seep into Jodi's consciousness, in the process transforming her from a rocker into a rockhound.

I admit I played a role in that unexpected development by dragging her (sometimes quite unwillingly—especially during those shows back in the early 1990s) to various fossil-centric events around the globe, places where my interest in trilobites continued to fester, and my collection continued to grow. However, rather than being continually bored by my arthropod obsessed activities, Jodi began to seek creative ways to occupy her time, as well as to make her own mark within this fossil-filled, male-dominated world. Working again with both a limited budget and limited knowledge, Jodi began the steady climb up the ammonite-lined stairway to cephalopod heaven. Show by show, and piece by piece—whether the piece in question was a Devonian ammonite from Madagascar or a Jurassic nautiloid from Germany—she quickly metamorphosed from a newbie to a dedicated hobbyist and then an expert in her field. This book is a direct result of that rather incredible transformation.

I take great pleasure in seeing how far Jodi has come over the last thirty or so years. She is now recognized as an ammonite authority by both her collecting peers and academicians—one of the primary reasons that she is now a field associate in paleontology with the world-renowned American Museum of Natural History. Where once the thought of spending an afternoon photographing fossil-filled rocks may have seemed no less than dreadful—especially after her exciting early career photographing the likes of Jon Bon Jovi and Eddie Van Halen—these days Jodi embraces such challenges with an unmatched zeal. Indeed, her enthusiasm toward continually expanding her ammonite collection has only been exceeded by her relish toward assembling the stellar contents of this book.

I freely admit that despite my affection for all things fossiliferous, ammonites tend to leave me somewhat cold; within my rather limited imagination, they do all kinda look the same. But I take solace in knowing that Jodi feels much the same way about my beloved trilobites, having once dismissively referred to them as "old sea cockroaches." But somehow we've managed to overlook such petty differences and appreciate our shared love of fossils. So when you're reading the carefully chosen words that fill this book and admiring the scores of beautiful color photos (virtually all of which Jodi took herself), realize that this tome isn't designed to be some stodgy, scientific effort done to garner academic renown within a publish-or-perish universe. Sure, there's plenty of science involved—Jodi wouldn't have it any other way. But this is the work of a passionate hobbyist, one who has assembled a world-class ammonite collection and now wants to share its contents, along with her invaluable (and often clever) insights on the subject, with an anxiously awaiting world. Ammonites are clearly among nature's greatest treasures, and with Jodi's enthusiastic help, we can all now better understand exactly why that is.

ANDY SECHER
Lifelong admirer and fellow fossil fanatic

# Foreword

Dr. Neil H. Landman

Ammonites, ancient sea creatures that lived millions of years ago, inspire wonder at the diversity of life in the distant past. With their hard calcareous shells, ammonites were perfect candidates for fossilization and, as a result, they occur as fossils around the planet, notably in North America, Madagascar, France, Japan, Russia, and even Antarctica. The shape of the ammonite shell varies widely from closely coiled spirals to, at the opposite end of the spectrum, long, straight cones. All these shells were subdivided into chambers by complex partitions so that the shell served as a buoyancy device, enabling the animal to stay afloat in the water column. The soft body of the animal, probably similar in morphology to that of a modern squid, was equipped with a siphon that allowed the ammonite to swim using jet propulsion. During the hundreds of millions of years that ammonites lived, they evolved a wide range of forms, with different species appearing at different times. Because of their rapid evolution, ammonites are regarded as ideal index fossils, with different species characterizing different time periods. For example, if you find an ammonite in a rock, you know instantly where you are in the geologic record—the Triassic, Jurassic, or Cretaceous. In this elegant book about ammonites, Jodi Summers masterfully covers all these topics and more, and her clear, readable text, accompanied by beautiful photographs, makes ammonites come alive again, providing a fascinating glimpse into an ancient world of wonders.

NEIL H. LANDMAN
Curator Emeritus
Department of Invertebrate Paleontology,
American Museum of Natural History

# WE DIG AMMONITES

***Porpoceras verticosum* Buckman 1914**
Order: Ammonitida. Appearance and lifespan: Early Jurassic, 182 to 176 MYA.
Location: Tournadous Section, Millau, Aveyron, Occitanie, France. Size: 7 cm.

# The Aristocratic Ammonite

## A New Wonder for the Shelf

It was near the dawn of the nineteenth century when the English-speaking world grew fascinated with ammonites. At that precarious moment in world history, the Napoleonic Wars were unsettling the European continent. In response, Britain's King George III encouraged his aristocracy to holiday close to home. Attempting to accommodate their ruler's advisory, the bourgeoisie flocked to Britain's coastal villages, charming sea-faring hamlets surrounded by frigid and foreboding waters. The weather was too cold for tanning, the seas too rough for swimming, and the rock-strewn jetties too jagged for strolling, yet these North Atlantic isles had their own distinctive charms. For recreation, the toney members of the privileged class searched for ammonite-filled nodules during low tide. These 200-million-year-old fossils were easy to find, and nearly everyone with the gumption to break a rock or two went home with a new treasure.

As word of mouth spread, collecting ammonite geodes soon became the au courant activity for refined ladies and gentlemen in search of something interesting to place in their cabinet of curiosities. On any given Sunday, you might have spied grandees in their top hats accompanied by ladies with velveteen shawls, all intently scanning the beaches of England's northeastern and southern coastlines for petrified prizes. The notorious high tides regularly battered the surrounding limestone cliffs, splaying an assortment of fossil geodes onto the beach. Low tide revealed a new hodgepodge of roundish concretions

Finding an ammonite pebble may make you smile.

*Dactylioceras* **Hyatt, 1867**
Order: Ammonitida. Appearance and lifespan: Early Jurassic, 182 to 175.6 MYA. Location: Saltwick Nab, United Kingdom. Size: 8 cm.

scattered along the shore. If you happened to pick a lucky rock, when it was gently split open with a geologic hammer, it might reveal a perfectly preserved spiral ammonite shell along with that creature's equally compelling negative impression. For many tourists, this fine fossil find made their holiday all that much more worthwhile. Now they had fantastic free souvenirs they could share with friends and loved ones, making it an inherently satisfying escapade for all involved. The pursuit of stone-encased fossil trophies along the English seaside soon established paleontology as an important new science for understanding early life on Earth.

"Ammonites are exquisite posterchildren for paleontology, evolution, and extinction, observes Kirk Johnson, Sant Director of the Smithsonian National Museum of Natural History in Washington, DC. "They lived everywhere in the ocean for a very long time and then they vanished along with the other iconic giants of the Mesozoic, the dinosaurs, in an event that happened in an instant. Ammonite fossils in their myriad species and many modes of preservation are staggeringly beautiful examples of the diversity of life on Earth."

## What Are Ammonites?

Ammonites are extinct creatures that swam in Earth's seas for about 350 million years. During that unimaginable stretch of planetary history, these marine invertebrates developed their own unique set of features, including buoyancy control and rapid evolution.

Biologically, Ammonoidea are molluscs as are a quarter of all aquatic organisms. Members of this phylum also include bivalves, gastropods, and squid. (FYI: There are approximately 85,000 extant mollusc species and an equal number of extinct variants.)

Some 66 million years later, ammonoids are still relatively easy to find. They typically resemble stone snail shells with an inverted whorl. A familiar shape, we see elements of their spiral design throughout nature—pinecones, flower petals, snails, and even the swirling shape of galaxies—to name a few. (It's been said that continuing the same shape through each successive whorl is an energy-efficient way for something to grow.) That's why the instinctual familiarity of an ammonoid's coiled shell makes them one of the most recognizable and pleasing of all fossil forms. "The spiral is one of the most ancient symbols in the history of humanity. It signifies the beginning and the end of life, the endless creation of one generation from the preceding one," noted Quirino Olivera, an archaeologist and the director of Peru's Montegrande Project.

Beyond being easy on the eye, these exceptional creatures serve a purpose and tell an amazing story in the process. Abundant predatory molluscs, this cephalopod subclass, were a squid-like, soft-bodied animals protected by a hard outer shell that grew continually throughout their lifetime. Species readily evolved, and later preserved, leaving their stories embedded in a stone matrix.

As the most prolific fossil in the geologic record, families have been discovered in sedimentary strata from Antarctica to Alaska. Scientists have already identified more than 10,000 genera of ammonoids, with new types still being named (and / or renamed) on a regular basis. During this cephalopod's nearly 400-million-year "swim," their conchs changed so rapidly that they now allow geologists to date a layer rock according to the species of cephalopod found in that stratum of stone.

A Western Interior Seaway ammonite with a well-preserved living chamber.

*Calycoceras tarrantense* **(Hyatt 1900) (sp)** split.
Order: Ammonitida. Appearance and lifespan: Late Cretaceous, 99.7 to 89.3 MYA. Location: Woodbine Formation, Tarrant County, Texas, United States. Size: 9.5 cm.

**AMMONITE BITE:** Technically speaking, "ammonite" is actually the colloquial term for Ammonoidea, a diverse subclass of cephalopods that emerged during the Devonian Period and died out as a result of the Chicxulub asteroid impact. The conversationally called ammonite refers to the Ammonitida, an order that lived from the Triassic through the dawn of the Paleocene. Older subclasses are known as ammonoids.

## What Makes Ammonoidea Special?

Ammonoids were among the earliest known creatures to swim above the sea bottom and venture into unexplored realms. Like most molluscs, the organism consisted of a soft-bodied animal protected by

**AMMONITE BITE:** The interior of the ammonoid shell between the protoconch and the body chamber is known as the phragmocone.

This nautiloid has a well-delineated central siphuncle. Notice how the septa curve away from the aperture.

*Cymatoceratidae* **Spath 1927 (sp)** split.
Order: Nautilida. Appearance and lifespan: Middle Jurassic, 164.7 to 161.2 MYA. Location: Ankirihitra, Ambato-Boeni, Boeny, Madagascar.
Size: 6.5 cm.

a firm outer shell. What made Ammonoidea and other early cephalopods remarkable was their ability to control their buoyancy.

Notice that the interior spiral conch is equipped with a series of compartments, which grow progressively larger toward the opening. The fleshy creature was always housed in the last and most prevalent body chamber. The other cavities were used for buoyancy. Collectively known as camerae, they are partitioned by concave support walls called septa. Each septum is attached to the outer conch at suture points. A tubelike siphuncle was positioned at the center of the venter, against the exterior wall of the conch. It regulated the amount of gas and fluids within the camerae to allow for near-neutral buoyancy.

Using this highly distinctive anatomical feature like a syphon, the ammonoid could add or dispel water from its chambers, allowing it to maintain nearly neutral buoyancy in epipelagic waters. Activating its hyponome to supply propulsion, likely using its arms to facilitate direction, the ammonoid thrust itself through the seas—shell first, creature last.

The ammonoid's innovative compartmental design deferred the hydrostatic pressure of water, giving the animal the ability to venture deeper into the ocean's niches. The structural septa seams are

**AMMONITE BITE:** Suture patterns are a significant factor in ammonoid species identification. All species have their own unique design, while family and genera have their own unique forms.

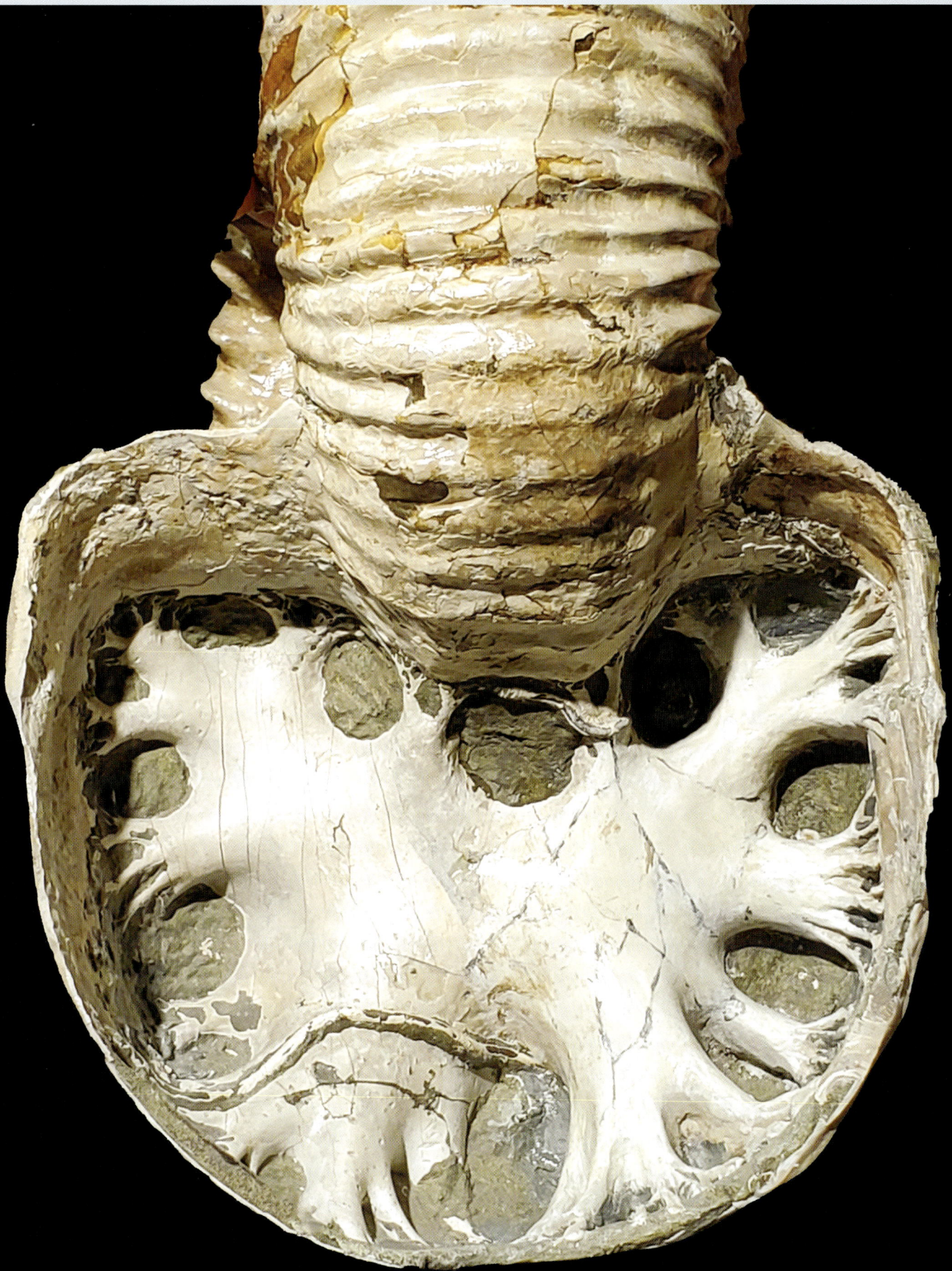

Ammonite septa construction is simple yet complex.

**Douvilleiceras inequinodum**
Order: Ammonitida. Appearance and lifespan: Early Cretaceous, 125.45 to 94.3 MYA. Location: Mahajanga Province, Madagascar.
Size: approximately 28 cm.

This *Cleoniceras* ammonite preserves the rainbow ammolite nacre while revealing the strengthening suture patterns that lie beneath the exterior.

***Cleoniceras* Parona and Bonarelli, 1897**
Order: Ammonitida. Appearance and lifespan: Early Cretaceous, 113 to 100.5 MYA. Location: Mitsinjo District, Boeny, Madagascar. Size: 5 cm.

visible below the protective nacre of the shell as suture lines. "As the ammonite grew larger and heavier, it would constantly be adding a new chamber wall at the backside of its mantle while adding more shell to the lip of the mantle at the opening or aperture," explained renowned cephalopod paleontologist Neal L. Larson.

## We Dig Old Ammonoids: Bundenbach, Germany

Although archaic by today's standards, in their heyday, ammonoids had more skills than most every creature in the ocean.

In southwestern Germany, there is a 405-million-year-old formation known as the Bundenbach Shale. It is one of the older preserved shelled cephalopod environments yet to be discovered. The Lower Devonian lithostratigraphic unit delineated as the Hunsrück Shale has yielded more than 260 animal species, including several early nautiloids

that had not evolved a coil. The Hunsrück Slate preservations have produced some of the most formidable stone captures on Earth. Intermingled in this extraordinary formation are pockets hiding detailed information about cephalopods—ammonoids, nautiloids and orthocerids—as well as trilobites, echinoderms, sponges, corals, brachiopods, jellyfish, gastropods, and worm trace fossils. Specimens with non-bio-mineralized body parts are dotted throughout this fossil-rich Lagerstätte—and not just invertebrates. The placoderm, a predatory armored fish, may have conserved soft-body material. These magnificent preservations required rapid burial in a low-oxygen sediment so that the decomposition of organic matter was significantly slowed.

Different areas of this formation highlight different biota. Brachiopods proliferated in the middle Kaub Formation in the vicinity of Gemünden. Concentrated around Bundenbach are trilobites with well-developed eyes, corals, and more than

A map of Germany with the region of the Hunsrück Formation location highlighted in red.

Image credit: The original uploader was ErnstA at German Wikipedia. Later versions were uploaded by Roßbacher at de.wikipedia., CC BY-SA 3.0, via Wikimedia Commons.

Switzerland, and of the Palaeontological Institute and Museum.

At more accommodating Hunsrück Slate environments in the Stuttgart region and in the vicinity of Wiesbaden, you can find several handsome, primitive cephalopod fossils. Included are the straight-shelled Orthoceras and several genera in the early stages of coiling. Waiting to be uncovered are genera like the open-whorled, prominently ribbed Anetoceras; the loosely coiled genus Ivoites; and the open, oval-shaped Gyroceratites; among others.

Extraordinary steps in discovering the secrets of the Hunsrück Slate were made in the 1970s. Wilhelm Stürmer, a chemical physicist and radiologist at the German engineering and technology conglomerate company Siemens, developed a new method to examine the slate fossils using X-rays. By manipulating medium-energy 25- to 40-kiloelectron-volt radiography, he created high-resolution movies and multidimensional images of unopened slates. His technology test revealed complex details of soft tissues that were previously not visible using conventional methods. Thanks in great part to Stürmer's examination, the Hunsrück Slate has been designated as a Konservat Lagerstätte because of the extraordinary preservation of soft tissue.

sixty distinct species of crinoids and starfish. Augmented by notable aggregations of sulfur and iron, Bundenbach fossils are sexy pyrite permineralizations upon dark, mud-gray slate. The downside for ammonoid fans is that the conditions disintegrated the calcareous shells of cephalopods as well as those of brachiopods and gastropods.

"Bundenbach is important because it has some of the oldest ammonoids, but similar forms of similar age occur also in Morocco and South China," noted Christian Klug, a member of the paleobiology department at the University of Zürich in

**AMMONITE BITE:** Konzentrat-Lagerstätte is a sedimentary deposit that exhibits an extraordinary volume of fossils. Known in English as concentration deposits, it's about quantity over quality. Conversely, Konservat-Lagerstätte is quality over quantity. Conservation deposits are known for the exceptional preservation of fossilized organisms, soft bodied parts, or trace fossils of scientific significance.

A close-up of Early Devonian agoniatites. Their shells are bowed, indicating the beginnings of the ammonoid's coil.

*Metabactrites fuchsi* **(De Baets et al., 2013)**
Order: Agoniatitida. Appearance and lifespan: Early Devonian, approximately 407.6 to 393.3 MYA.
Location: Kaub Formation, Bundenbach, Germany Size: 9 cm.

Image credit: Courtesy of Dr. Christian Klug.

# The Legalities of Fossil Prospecting in Germany

It is often difficult to find the fossil-collecting laws of foreign countries published in English. Fortunately, there are online forums that discuss these rules for many parts of the globe. Locals are generally proud to share their knowledge and offer relatively accurate information on digging for fossils in their country and then exporting them.

Per Germany's current cultural property protection laws (KGSG), fossils are often not considered cultural assets. A translation of the cooperative assessment by the paleontological societies of Germany, Austria, and Switzerland in November 2015 observed that "90 percent of all fossil discoveries have little or no scientific or commercial value." These specimens are not considered to be part of their home-land's "cultural heritage." The takeaway—apart from a few outstanding preservations—is that German law does not consider fossils to be "cultural property." The same rules apply to minerals or geological samples.

To be on the safe side, it has been suggested by European collectors that fossil enthusiasts traveling out of Germany carry a printout of the *Revised Paleontology Background Paper of the Federal Government* to show customs, if necessary.

A mélange of ammonoid species.

## Many Faces

Like trees in a forest or like a crowd of humans, ammonoids tended to look similar yet different. Except for early and late varieties, Ammonoidea were typically spiral in form, but they varied significantly in size. Shell diameters ranged from less than 1 centimeter to more than 2 meters (¼ inch to more than 6.6 feet). These unique marine creatures varied not only in circumference but also in thickness, ornamentation, and number of whorls. Conchs evolved from simple spirals to elaborate tests adorned with spines, bullae, nodes and constrictions. The most extraordinary phase of ammonite development was toward the end of their run; during the Cretaceous Period, these cephalopods radically changed shape. "Ammonites are intriguing; their spectacular shells make them interesting; that spiral shape draws people in," observed Joshua Slattery of the Department of Physics at the University of North Florida.

The soft-bodied animal looked somewhat like a cuttlefish in a shell. It had eyes and arm appendages and was an inimitable hybrid of three of today's four living cephalopod classmates—the squid, nautilus, and the cuttlefish. (The octopus is quite different in many respects.) From the Paleozoic to beyond the Mesozoic, ammonoids inhabited the seas with a cast of creatures ranging from trilobites to mosasaurs. They became collateral damage when the K-Pg (Cretaceous–Paleogene) extinction event took out 80 percent of life on Earth, including all nonavian dinosaurs.

## The Debut

Paleozoic seas were a roller-coaster ride through changing environments. Oxygen-deprived stages caused the extinction and rapid evolution of many marine species. One of the new groups that developed after a Cambrian Period elimination became Ammonoidea. Researchers combing for fossils around Bacon Cove in eastern Newfoundland, Canada, believe they have discovered microfossils indicating that the subclass may have been around as early as 522 million years ago. "That would mean that cephalopods emerged at the very beginning of the evolution of multicellular organisms during the Cambrian explosion," according to Anne Hildenbrand, a researcher at the University of Heidelberg.

These early predators had a simple, tapered shell and were rather mundane looking on the outside, but the chambered conch was unlike anything that had yet to evolve on Earth. Effective hunters, these newly minted cephalopods used their many arms to capture small prey such as jellyfish, trilobites, worms, brachiopods, and the like. Evolution produced the family *Ellesmeroceratidae*, which proliferated in the Early Ordovician, yielding some eight new cephalopod orders that filled diverse ecological lifestyles . . . Then the seas changed.

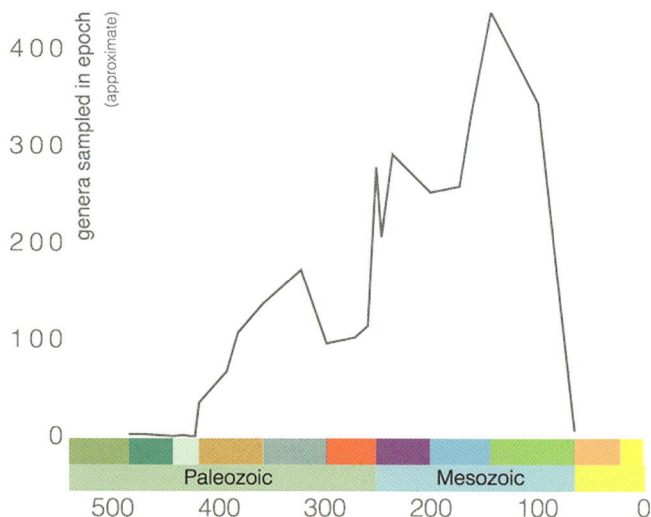

Genera level diversity of Ammonoidea during their time on Earth.

Image credit: Jonathan R. Hendricks, Phanerozoic genus level diversity of Ammonoidea (graph generated using the Paleobiology Database Navigator). Creative Commons Attribution-Noncommercial-ShareAlike 4.0 International License.

When the oceans are choked of oxygen content and life is struggling to survive the environmental conditions, that's an anoxic event. The reasons and results are compound and varied. Although now a rare occurrence, ocean oxygen content affected ammonoid evolution throughout their time on Earth. Instances were associated with brutal climate change, often caused by tectonic plate movement and extensive volcanic activity, bolide impacts, as well as noxious algae blooms. It has been suggested that anoxic events caused marine attrition during Ordovician–Silurian, Late Devonian, Permian–Triassic, and Triassic–Jurassic extinctions and contributed to a number of lesser annihilation events.

As the Ordovician slid into the Silurian Period some 444 million years ago, conditions on Earth created a series of toxic conditions that challenged the development of life. The causes for these eliminations are complex and multilayered, involving landmass and climate shifts and challenging maritime conditions. As continents slowly repositioned themselves, bursts of volcanic activity and tectonic uplifts drained epicontinental seas. In turn, this activity increased ocean salinity concentrations and altered currents and oxygen intensities. Newly minted terrestrial plants and phytoplankton sucked up the carbon dioxide, exacerbating fluctuations of atmospheric and oceanic CO2 intensities. Marine animals saw their habitats cool and shrink dramatically. Middle latitude tropical greenhouse conditions were undercut by boreal bottom currents. This situation resulted in an ice age that locked up ocean water, forcing global sea levels to drop by more than 100 feet. For more than five million years, the lore of the seas was *adapt or die.* By the conclusion of this Ordovician-Silurian extinction, Earth had lost its majority of trilobites, corals, crinoids, brachiopods; fifty-five out of one hundred genera of nautiloids disappeared. In all, the die-off destroyed an estimated 85 percent of all Ordovician species.

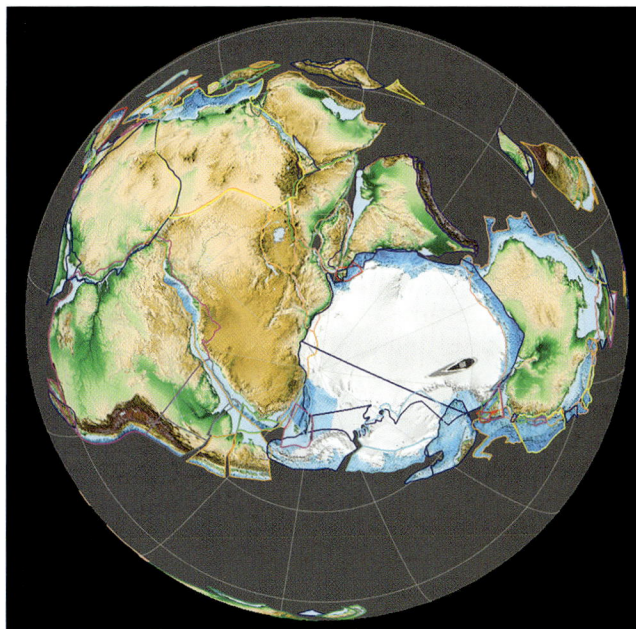

A spherical map of Gondwana: the world 420 million years ago as seen from the South Pole.

Image credit: Fama Clamosa, CC BY-SA 4.0, via Wikimedia Commons.

Once oxygen replenished the seas, the renewed ecosystems of the Silurian Period were remarkably similar to what had existed in the Ordovician. Nautiloids, brachiopods, cephalopods, bryozoans, and trilobites again flourished, becoming even more widely distributed than their predecessors. But there was also expansion. A variety of early cephalopods appeared. Species like *Pomerantsoceras pollux* and *Lituites lituus* emerged and quickly adopted distinct ecological niches. The Silurian saw the undersea diversification of jawed and bony fish, and terrestrial multicellular plants took root. As the Devonian Period dawned—heralding trumpets, please—we have the first ammonoid: *Agoniatitida.*

**AMMONITE BITE:** Agoniatitida is the ancestral order for the entire Ammonoidea subclass.

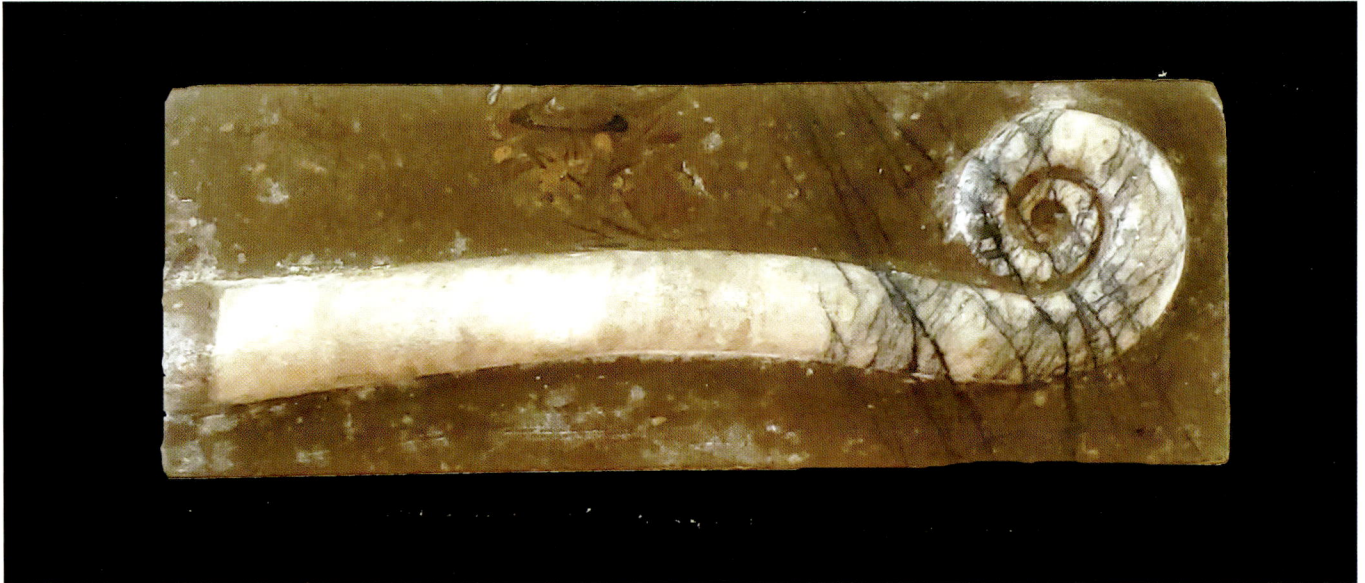

*Lituites* **Bertrand, 1763**
Order: Orthocerida. Appearance and lifespan: Middle Ordovician, 463.5 to 418.7 MYA. Location: Baota Formation, Songtao Miao Autonomous County, Guizhou Province, China. Size: 12 cm.

## Coil Confirmed

With the exception of orthocones and eccentrically shaped heteromorphs, the vast majority of ammonoids developed a symmetrical, planispiral form. Shell sizes varied from smaller than a nail head to larger than a tractor tire. These unique marine creatures had conchs that not only differed in size but also in thickness, ornamentation, number of visible whorls, and the suture patterns. A watermark of sorts, these lines denote where and how the septa attaches to the phragmocone. The structure strengthened the shells and made them increasingly impervious to water pressure. The suture patterns are unique to each genus.

During their more than 350 million years in the world's seas, ammonoid shells evolved from small simple spirals built of boxlike compartments to hydrodynamic artforms. By the time this cephalopod subclass went extinct some 66 million years ago, carapaces had evolved into corkscrew helixes, trombone-like forms, multidimensional squares, and other exotic shapes.

*Anetoceras* **Schindewolf, 1935 (sp)**
Order: Agoniatitida. Appearance and lifespan: Early Devonian, 409.1 to 402.5 MYA. Location: Bundenbach—Eschenbach, Ostalbkreis, Stuttgart Region, Baden-Württemberg, Germany. Size: 2.5 cm.

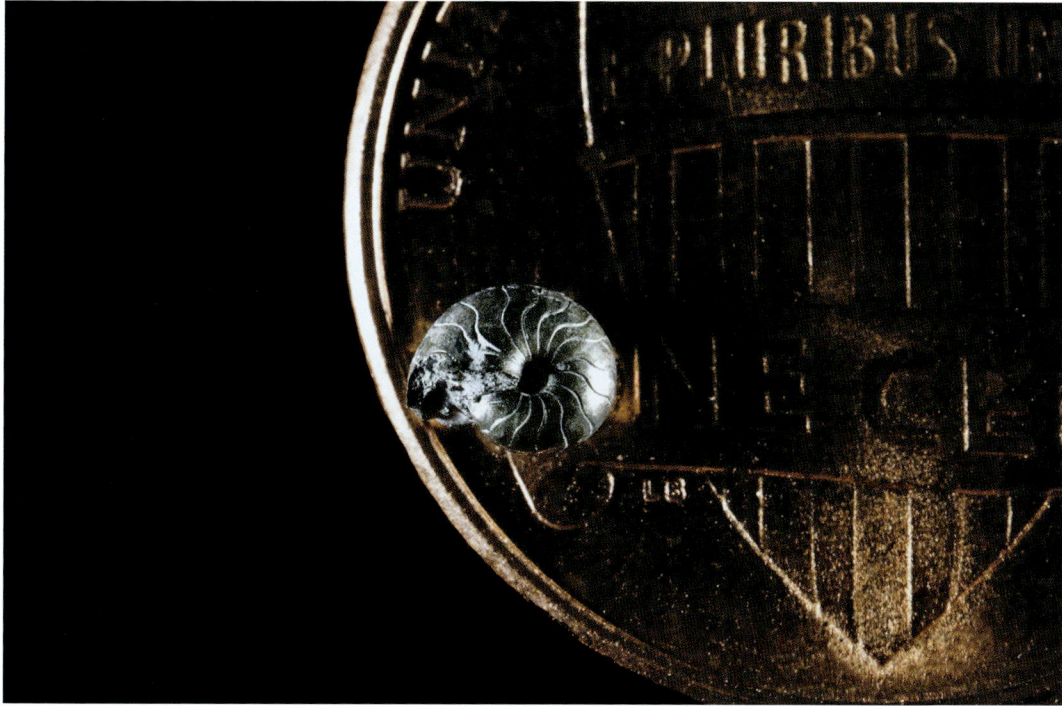

***Agoniatites vanuxemi* (Hall, 1879)**
Order: Agoniatitida. Appearance and lifespan: Middle Devonian, 393.3 to 387.7 MYA. Location: Widder Formation, Southwest Ontario, Canada. Size: 1 cm.

Image credit: Mike Meacher, Stormbed Paleontological.

## We Dig Old Ammonoids: New York, United States

Academicians, politicians, scientists encourage individuals to collect some of the oldest round ammonoids—the ones that dot the Marcellus Shale Formation in western New York State. This vast, rich Devonian terrain is spectacularly scenic—like the purple mountain majesties from the anthem "America the Beautiful." An extraordinary, little-explored preservation, it hides early Ammonoidea. Countless streams and revealing roadside curb cuts offer collecting possibilities that go on for counties. Find the cephalopod-profuse Cherry Valley Limestone Lagerstätte and you may discover discrete beds of *Agoniatites vanuxemi*—a precious Early Devonian ammonoid.

Tens of thousands of shells are embedded in the black shale layer; this profusion of fossil treasures tells a bittersweet tale. The mass preservation may indicate that numerous nautiloids and newly emerged ammonoids were struggling to survive in a changing marine environment—and clearly didn't succeed.

"The Cherry Valley apparently reflects deposition in shallow, calm waters, affected by cyclic variation in delivery of clastics, possibly combined with sea level fluctuations," noted John Cottrell, a paleogeologist at the University of Rochester. "Circulation and oxygenation were effective enough to support relatively diverse benthic fauna."

The Devonian massif traverses southwestern New York State like a tilted stack of books, capturing a tumultuous tale of world development.

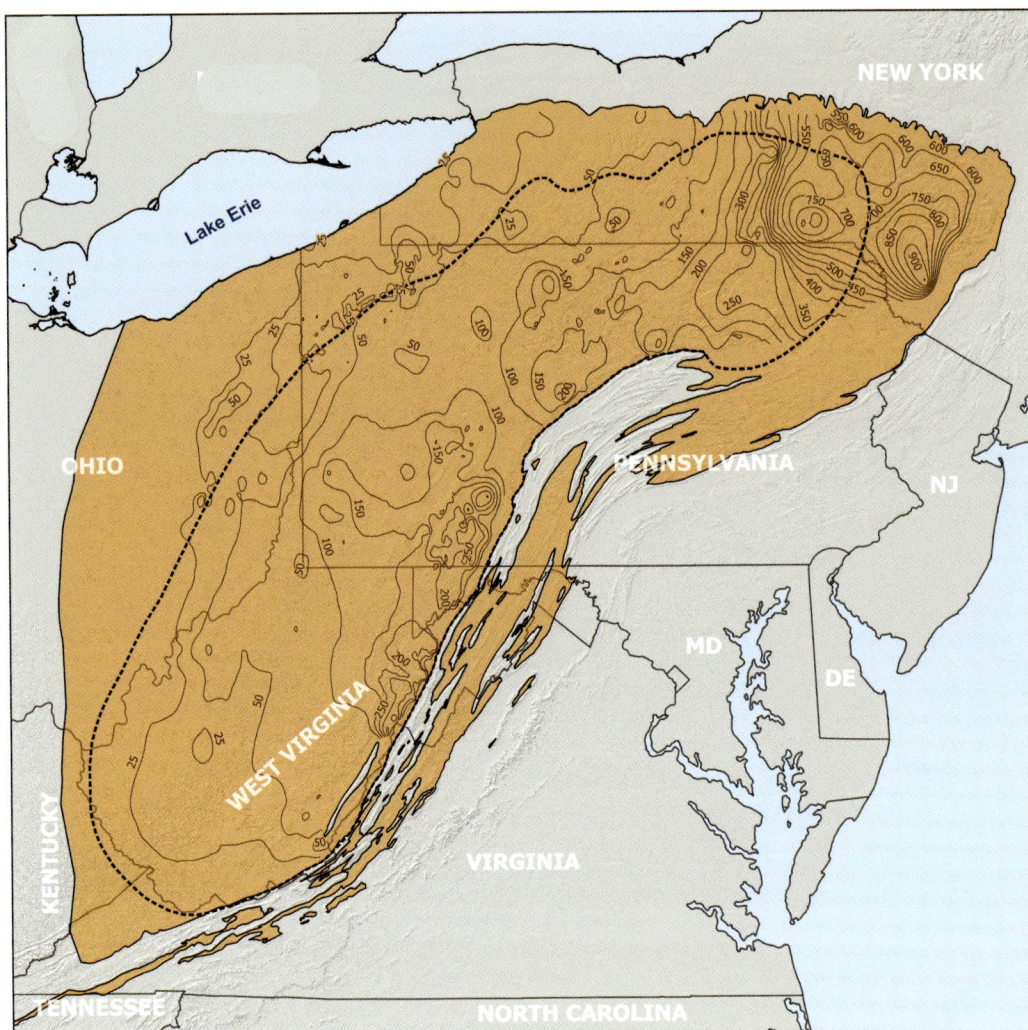

The Marcellus Shale Formation spans several northeastern states. This isopach map displays the stratigraphic depth between an upper and lower horizon.

Image credit: Gretarsson, CC BY-SA 4.0, via Wikimedia Commons.

This substantial deposition consists of elongated belts of folded, thrust, and faulted marine sedimentary rocks and volcanic knobbles interspersed with fragments of ancient ocean floor. These slices of sea bottom were contorted by the continental drift of tectonic plates and make for excellent ammonoid collecting locations.

For this excursion, local paleontologists are the best guides. Browse the rockhounding forums for paleo clubs, area prospectors, digging locations, and other needed details. Know where you're going because the black Marcellus Shale extends through the Appalachian Basin for hundreds of miles. In addition to New York State, the massif courses through Pennsylvania, Ohio, Maryland, and West Virginia.

A typical Marcellus Shale outcrop.

Image credit: Dhaluza, CC BY-SA 3.0, via Wikimedia Commons.

## Black Shale Occurrences

The darkest, most fossiliferous shale layers typically indicate a mass death event. The metamorphic black rock interval captured in the Marcellus Shale Formation coincides with a major change in the world's oceans known as the Kačák Event. This marine extinction went down approximately 387 million years ago, when tectonic plate movement agitated the ocean floor. This occurrence led to rising sea levels, which in turn created an assortment of deep, stagnant low-oxygen pools within the larger marine basins. When geological processes kicked up the anoxic water, it rose to the surface, slowly asphyxiating whatever unfortunate creatures happened to be in the vicinity. The ebb and flow of this series of mass mortality conditions deposited shales with a rich concentration of common and rare cephalopods intermingled with skeletal limestones. "Perhaps the Marcellus Shale captures a series of mass mortality events suffered by migrating populations during times of sea level rise," offered an area rock hound. "That could account for the cephalopod clusters which are exclusive to specific layers in particular sections."

Avocational and professional paleontologists have been collecting Cherry Valley Formation specimens the since the mid-nineteenth century; yet academics realize there is still much to be discovered about this extraordinary formation of ancient ocean life. Too many New York cephalopod species are known from only one or a few specimens and exclusive pockets of unique creatures. There's so much more of this preservation to be revealed. Area scientists invite you to contribute to history and join the search.

The area's freezing winters cause erosion, freeing selections of rocks and fossils around creeks and roadcuts that can be prospected after the spring thaw. "The fossils weather out of the banks where the shale is exposed. The best time to look is after a good rain," elaborated Mike Meacher of Stormbed Paleontological. (His focus is on the Arkona Shale, on autonomously owned land on the Canadian side of the formation, north of the Lake Erie shoreline.) "They are found by looking very closely at the exposed clay with your nose almost to the ground as they are quite small and 'stick' to the bank exposure. When you find them, they are very reddish from oxidation and often incomplete. Other tiny pyrite fossils are found in the same way at this spot, including brachiopods; tentaculites, a mysterious genus of conical fossils; crinoid bits; et cetera."

Unless you can train your dog to spot these peppercorn-sized fossils, you will have to plan on being on your hands and knees in a cool, damp environment looking for extraordinarily small things. It's a memorable experience, especially when you find a pocket and come away with a grouping of these conserved gems. "I prepare them with air abrasion at low pressure and either hold on to them tightly between my fingers or stick them into plaster so I can blast each side with dolomite powder," shared Meacher about readying the finds from this unique Devonian conservation for display.

As you discover new treasures, look for the telltale wavy suture pattern of the goniatite. And if think you found something special, geotag the location and please do share your fossil with the New York Paleontological Society. They may document and record the details of your find using their impressive spectroscopy and X-ray diffraction equipment, boosting the provenance to your new treasure.

*Manticoceras* **Hyatt, 1884**
Order: Agoniatitida. Appearance and lifespan: Late Devonian, 382.4 to 379.5 MYA. Location: Cashaqua Shale, Ontario County, New York. Size: 6 cm.

**AMMONITE BITE:** Partial preservation of cephalopods is a noted characteristic of the Cherry Valley fossilizations. They were apparently rapidly buried as many topsides became worn and battered in agitated seas.

Just as the Cherry Valley Limestone Lagerstätte is a sweet spot for *Agoniatites*, different niches of the Marcellus Shale tell various intimate stories about ammonoid history and survival.

- In a remote corner of southwestern New York State, there is a sparse sandstone and seabed conglomerate where paleontologists have recovered one *Maeneceras*, the youngest goniatite found in the state. Because it was part of an abundant genus in Morocco, this lone specimen is believed to be postmortem traveler that floated across the Afro-Appalachian seaway link at a time when most of the planet's landmasses were gathered in the Southern Hemisphere.

- The ammonoid species *Prochorites alveolatus*—with its large rounded lateral lobe and concave outer rim—were ocean drifters. Their presence in the Marcellus Shale indicates that North America once had a seaway connection with Australia.

- *Cheiloceras amblylobum*, known as an index species from Germany, has so far been discovered in just a single New York shale bed.

- The cosmopolitan genus *Pharciceras* is known in the Empire State from the iconic Tully Formation.

## Tully Formation

South of Syracuse is the youngest Devonian limestone layer in the State of New York. It pokes through the array of detrital sedimentary rock walls and rubble that make up the Catskill Delta. Known for its shelly fauna of corals and brachiopods, this location records a global extinction that took place approximately 385 million years ago. It represents a significant early pause in ammonoid evolution, and bitterly, it proved to be the undoing of the Agoniatitida ammonoid order. Sweetly, it allowed for the rise of a burgeoning breed of ammonoid—the goniatites.

*Protornoceras polonicum* **Dybczynski, 1913**
Order: Goniatitida. Appearance and lifespan: Late Devonian, 376.1 to 365 MYA. Location: Kowala, Kielce County, Świętokrzyskie Voivodeship, Poland. Size: 1.5 cm.

# The Legalities of Fossil Prospecting on Public Lands in the United States

The Bureau of Land Management (BLM) is the government agency responsible for administering federal lands in the United States. This agency, part of the Department of the Interior, oversees more than 247.3 million acres (1,001,000 km²), more than 10 percent of the country's overall landmass. When it comes to you, me, and McGyver collecting fossils on federal land, the rules state that "You may collect reasonable quantities of common invertebrate fossils such as molluscs and trilobites, but this must be for personal use, and the fossils may not be bartered or sold." Ammonoids are normally considered common invertebrate molluscs, so you have ample flexibility as a cephalopod collector searching on public lands.

The guidelines are quite different for vertebrates. "Only accredited researchers with a BLM permit are allowed to collect vertebrate fossils such as dinosaurs, mammals, fish and reptiles, as well as uncommon invertebrate or plant fossils. Collected fossils remain public property and are placed with museums, universities, or other public institutions for study and exhibition." While we're reviewing the rules for prospecting public lands, resources found in caves, including plant, animal, and geologic features, are all federally protected and "may not be altered, damaged, or removed."

*Gundolficeras bilobatum* **Becker 1995**
Order: Goniatitida. Appearance and lifespan: Late Devonian, 376.1 to 359 MYA. Location: Kowala, Kielce County, Świętokrzyskie Voivodeship, Poland. Size: 1.5 cm.

The cephalopod display at the Denver Museum of Nature and Science highlights the extraordinary variety of ammonites.

Straight-shelled baculites from center top to bottom left:

***Baculites* Lamarck, 1799.** Order: Ammonitida. Family: Baculitidae. Appearance and lifespan: Late Cretaceous, 70.6 to 66 MYA. Location: Pierre Shale, Butte County, South Dakota, United States.

***Baculites grandis* Hall and Meek, 1855.** Order: Ammonitida. Family: Baculitidae. Appearance and lifespan: Late Cretaceous, 72.1 to 66 MYA. Location: Pierre Shale, El Paso County, Colorado, United States.

Baculite cluster:

***Baculites compressus* Say, 1920.** Order: Ammonitida. Family: Baculitidae. Appearance and lifespan: Late Cretaceous Period, 83.5 to 70.6 MYA. Location: Pierre Shale, Pennington County, South Dakota, United States.

Large single planispiral ammonite in the center:

***Mortoniceras drakei* (Young, 1957).** Order: Ammonitida. Family: Brancoceratidae. Appearance and lifespan: Early Cretaceous, approximately 113 to 105, MYA. Location: Duck Creek Formation, Pecos County, Texas, United States.

Planispiral ammonite clusters left to right:

***Dactylioceras athleticum* (Simpson, 1855).** Order: Ammonitida. Family: Dactylioceratidae. Appearance and lifespan: Early Jurassic, 191 to 174 MYA. Location: Schlaifhausen, Germany.

***Promicroceras planicosta* (Sowerby, 1814).** Order: Ammonitida. Family: Eoderoceratidae.

***Asteroceras obtusum* (Sowerby, 1817).** Order: Ammonitida. Family: Asteroceratidae. Appearance and lifespan: Early Jurassic, 196.5 to 189.6 MYA. Location: Lower Lias Formation, Dorset, England, United Kingdom.

***Prionocyclus wyomingensis* Meek, 1876.** Order: Ammonitida. Family: Collignoniceratidae. Appearance and lifespan: Late Cretaceous, 93.5 to 89.3 MYA. Location: Frontier Formation, Albany County, Wyoming, United States.

Heteromorph ammonites from left to right:

***Hyphantoceras* Hyatt, 1900.** Order: Ammonitida. Family: Nostoceratidae. Appearance and lifespan: Late Cretaceous, 101 to 66 MYA. Location: Westphalia, Germany.

***Didymoceras* Hyatt, 1894.** Order: Ammonitida. Family: Nostoceratidae. Appearance and lifespan: Late Cretaceous, 83.5 to 70.6 MYA. Location: Pierre Shale, Pueblo County, Colorado, United States.

***Oxybeloceras crassum* (Whitfield, 1877).** Order: Ammonitida. Family: Diplomoceratidae. Appearance and lifespan: Late Cretaceous, 75 MYA. Location: Pierre Shale, Weston County, Wyoming, United States.

# 2

# Orders and Disorder

## Conflicting Points of View

The eras-long transformation of this coiled cephalopod was, in major part, based on extinction survivors adapting to new conditions. With each elimination event, the most vulnerable creatures disappeared, and later, new life forms arrived in their place. Some orders survived for tens of millions of years; others barely made it a mega-annum. The regular ebb and flow of ammonite families has led to an ongoing discourse, debate, and consternation within the academic community. The last time the science world agreed on how to organize and categorize the extraordinary variety of ammonoids was back in the nineteenth century. The defining moment came when renowned paleontologist, Karl Alfred von Zittel, director of the Natural History Museum of Munich, introduced the subclass term Ammonoidea. He did so in *Handbuch der Palaeontologie*, one of the era's rare environmental encyclopedias. The professor's work was then augmented by an American Civil War captain and noted academic, Alpheus Hyatt, who established the orders Goniatitida and Ceratitida. He noted, "The remarkable sudden appearance and fully developed structures of these earlier ammonoids finely illustrates the fan-like character of the evolution of forms from chronologic center of distribution and the quickness with they must have spread and filled up the unoccupied habitats." (Hyatt's studies led to the founding of the Woods Hole Marine Biological Laboratory, still an international center for research and education.)

In the twentieth century, ammonoid identification became whorls of chaos. Global paleontologists have been able to agree on only three orders of ammonoids: Goniatitida, Ceratitida, and Ammonitida. But since this baseline was established, specialists have agreed to disagree. "Species are open nomenclature," confirmed Neil Landman, curator emeritus, Fossil Invertebrates, Division of Paleontology, Richard Gilder

Graduate School of the American Museum of Natural History.

The debut edition of the venerable "Geological Society of America's Treatise on Invertebrate Paleontology" (Part L, 1957) declared that what we know as the cephalopod subclass Ammonoidea was simply an order with eight suborders. (Perhaps it was a typo? Subsequent editions raised Ammonoidea to a subclass and elevated most of the suborders to orders.) "Most collectors and academics consider the revered treatise to be a revered relic of a bygone era," groused members of the scientific community. "Recent publications go into different areas into different numbers. And so the debate rages on."

While the United States was clarifying classifications, Comrade Professor V. E. Ruzhencev of the Paleontological Institute of the Academy of Sciences of the Union of Soviet Socialist Republics in Moscow released a paper listing five orders of ammonoids. Common information sources in the United States site six orders, and there is debate about whether to include others, and which others. "Devonian to Triassic ammonoids were usually considered to be a rank above order (with the orders Goniatitida, Clymeniida, Ceratitida, etc.), while Jurassic and Cretaceous ammonoids were ranked lower, (with Ammonitina and Phylloceratina as suborders)," concluded an esteemed international team of paleontologists. That group cited seven ammonoid orders, raising the suborder Phylloceratina to the order Phylloceratida. Academics in the United States support additional recognition for suborders Ancyloceratina and Lytoceratina.

Let's dive into the differences among the six accepted ammonoid orders—Agoniatitida, Goniatitida, Clymeniida, Prolecanitida, Ceratitida, and Ammonitida. Additionally, let us acknowledge three pivotal groups—Phylloceratina, Ancyloceratina, and Lytoceratina—all of which appeared in the Jurassic and survived into the Late Cretaceous—and sometimes a bit beyond.

## Agoniatitida

**First Recorded Appearance: Early Devonian, ±405 million years ago**

**Last Recorded Appearance: Middle Devonian, ±385 million years ago**

Evolution led to ammonoids some 405 million years ago, when Agoniatitida debuted. The earliest member of the cephalopod subclass Ammonoidea straddled the line between old and new. Agoniatitida inherited characteristics from its ancient uncles, the Orthocerida, Bactritida, and Nautiloidea, but it had some completely original aspects that made it new, noteworthy, and different. Shaped a bit like a really small blueberry, it had achieved the coiled shell we have come to appreciate with this fauna. Inside, Agoniatitida

*Agoniatites vanuxemi* **(Hall, 1879)**
Order: Agoniatitida. Appearance and lifespan: Middle Devonian, 393.3 to 387.7 MYA. Location: Widder Formation, Southwest Ontario, Canada. Size: 1 cm.

was still a design in process. Its chambers had a retro pattern—not unlike the nautiloid's backward facing septa. To aid in buoyancy, there was a tubular siphuncle that listed toward the center of each whorl of the creature's shell. The suture seams that separated its chambers had simple undulations, like the sea on a calm day.

An innovative cephalopod order, Agoniatitida appeared during the Devonian Period. These creatures existed for about 30 million years, when they were abruptly wiped out during the environmental discord of the Late Devonian Kellwasser Event—one of several distinct pulses that unfolded over several hundred thousand years, resulting in a planetary transition of pelagic and reef biota. Fossilized specimens have been discovered in Africa, Asia, Australia, Europe, and North America.

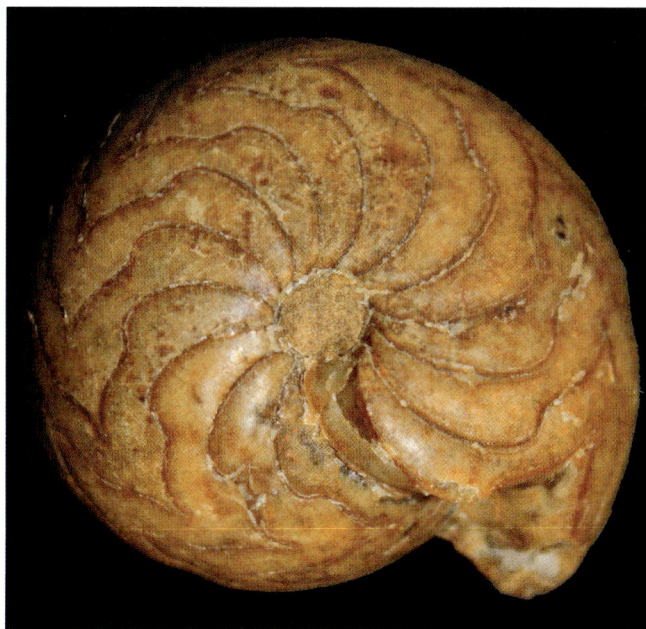

**Goniatites multiliratus (Chocktawensis) Gordon, 1962** (sp)
Order: Goniatitida. Appearance and lifespan: Middle Carboniferous, 336 to 326.4 MYA. Location: Johnston County, Oklahoma, United States. Size: 4 cm.

## Goniatitida

**First Recorded Appearance: Middle Devonian, ±393 MYA**

**Last Recorded Appearance: Late Permian, ±251 MYA**

Known to their admirers as goniatites, the superorder Goniatitida sprang from the more primitive Agoniatitida on an endless summer's night about 390 million years ago in the wake of the Choteč Event. These Paleozoic ammonoids evolved at a time when water levels were rising and ocean oxygen was increasing. Larger than their predecessors, they were characterized by simple suture patterns of irregular zigzags or gentle hilly patterns that connected the interior camerae, creating a sturdy infrastructure for the growing shell. "Mechanical functional interpretations generally concern stress resistance, with more complex divider walls acting as buttresses," supported David J. Peterman of the Department of Geology and Geophysics, University of Utah.

Like all ammonoids, the goniatite animal lived in the largest chamber at the front of its external conch. New and improved, for the first time ever, the all-important siphuncle had migrated to the inner surface of the outer shell of the ammonoid, which would prove to be its most efficient location. This placement accelerated the animal's hydrostatic durability, allowing goniatite shells to increase in size over tiny predecessors. Research shows that the conch grew up to 80 centimeters (31.5 inches) in diameter and could reach 5 centimeters (2 inches) across. The shell's coiling pattern was predictable—evolute and sleek, or involute and bulbous.

Occasional species had bumps, spines, ribs, or other forms of ornamentation, but what made goniatite conchs special were their more complex suture schematic. "The septal pattern underneath the overlying exterior shell of the

***Gonioclymenia*** **Hyatt, 1884**
Order: Clymeniida. Appearance and lifespan: Late Devonian, 364.7 to 360.7 MYA.
Location: Errachidia Province, Drâa-Tafilalet Region, Morocco. Size: 21 cm.

*Clymenia laevigata* **(Muenster, 1832)**
Order: Clymeniida. Appearance and lifespan: Late Devonian, 364.7 to 360.7 MYA. Location: Kowala, Poland. Size: 3 cm.

ammonoids provided superior strength to the overlying shell by providing a complex underlying support system," observed Neal L. Larson. "They likely did not dive very deeply," elaborated Christian Klug. "Bear in mind that all the forms we know are actually from epicontinental seas rather than oceans, so water depth rarely exceeded 200 meters."

Recent research and three-dimensional modeling techniques have allowed academics to extrapolate on the possible benefits of complex suture patterns. Some studies indicate they may have served as a defensive trait. Others suggest that the complex ammonite seams held water, allowing for enhanced buoyancy control. "They may also have enabled better balance, bigger size and external shapes that favor speed," offered Kathleen A. Ritterbush of the Department of Geology and Geophysics at the University of Utah.

An ideal index fossil, various goniatite genera were around for some 140 million years. They survived the Late Devonian Period extinction events to flourish during Carboniferous before calling it a day at the end of the Permian Period.

## Clymeniida

**First Recorded Appearance: Middle Devonian, ±375 MYA**

**Last Recorded Appearance: Late Devonian, ±360 MYA**

Sometimes referred to as Clymeniina, this order ranged from Europe to Africa and Australia.

Fossils reveal a unique dorsal siphuncle that ran along the inside of the coils. Clymeniida's shells were as large as 80 centimeters (31.5 inches) in diameter, but they had variety. The whorl could be evolute or involute and might grow ribs or spines as it matured. The species that are most recognizable were tightly wound like a rope, similar to Clymenia laevigata.

"The theme of early 'experimentation' shows up amidst Devonian ammonoid diversity. The clymeniids constitute an arch example, with their siphuncle on the opposite side of the body from what proved to be the 'normal' ammonoid condition," observed Niles Eldredge, emeritus curator, Division of the Paleontology at the American Museum of Natural History. It was "an experiment that failed to survive the late Devonian biotic crisis, thus forever depleting ammonoid morphological diversity."

> **AMMONITE BITE:** Ammonoid shells may be evolute with inner whorls exposed, or they can be involute with only the outermost whorl showing.

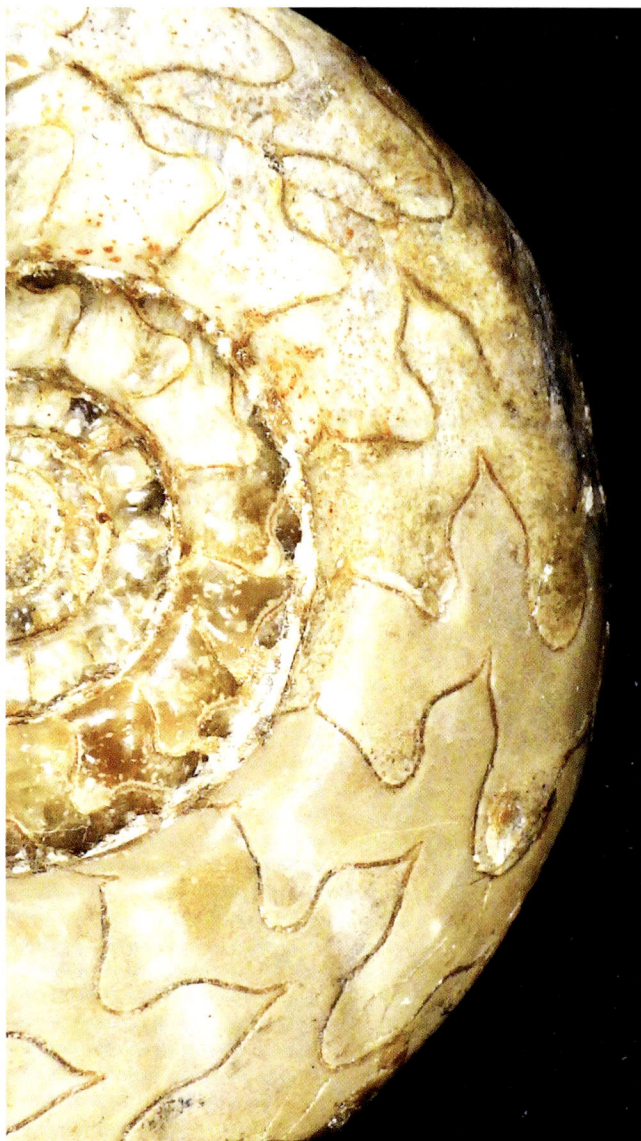

***Propinacoceras (Uraloceras) aktubense* Ruzhentsev, 1939**
Order: Prolecanitida. Appearance and lifespan: Permian, 290.1 to 279.5 MYA. Location: Zhil-Tau Hill, Aktyubinsk Region, Kazakhstan. Size: 3.5 cm.

## Prolecanitida

**First Recorded Appearance: Late Devonian, ±365 MYA**

**Last Recorded Appearance: Early Triassic, ±250 MYA**

A small and stable order, the Prolecanitida survived about 108 million years and generated forty-three named genera and about 1,250 species. The suture patterns of this group exhibit evolved complexity; peaks and valleys are more convoluted, which strengthened the shell. They also had curiously short living chambers. A work in progress, its siphuncle had migrated to its more evolved position along the ventral margin, but, like the nautiloid, its design was retrochoanitic, with supporting septa extended toward the apex. The order Prolecanitida survived through the Great Dying at the end of the Devonian Period. This order was able to endure and thus provided the root stock from which all later Mesozoic ammonoids were derived.

## Ceratitida

**First Recorded Appearance: Permian, ±279 MYA**

**Last Recorded Appearance: Middle Jurassic, ±165 MYA**

The unique saddle and lobe suture design of Ceratitida is characteristic of this order. Ceratitida overwhelmingly produced planispiral, coiled discus-shaped shells that may or may not have revealed their whorls. The order displayed notable advancements since Agoniatitida. Among its assets was its more complicated suture pattern, which allowed the shell to deflect increased water pressure and avert the bite of a predator attempting to crush the conch. In addition, the phragmocone showed the sophisticated ventral location of its siphuncle, which was a more efficient position.

*Ceratites* de Haan, 1825
Order: Ceratitida. Appearance and lifespan: Middle Triassic, 242.0 to 235.0 MYA. Location: Schweinfurt, Bavaria, Germany. Size: 10 cm.

## Ammonitida

**First Recorded Appearance: Early Triassic, ±247 MYA**
**Last Recorded Appearance: Paleocene, ±65 MYA**

These ammonites are the classic cephalopods we have come to know and appreciate. Treasured for their size, shape, and elaborate feathered suture structure, they are appreciated the world over for their decorative value. Families of this order evolved swiftly; a few stayed around for the duration, while other offshoot species existed for less than a half million years. Their rapid radiation makes Ammonoidea significant for biostratigraphy because they simplify the process of dating the matrix around them.

Ammonitida was the ultimate order in the Mesozoic. "Ammonites were first subdivided at generic level on the basis of shell coiling, heteromorphs or 'uncoiled' shells being separated after 1799 from the normally coiled shells," observed D. T. Donovan of the Department of Geological Sciences, University College London. "The latter were referred to a single genus, Ammonites, until the late nineteenth century. After the publication of Charles Darwin's

*Douvilleiceras inaequinodum (Douvilleiceras orbignyi)*
**Quenstedt 1849**
Order: Ammonitida. Appearance and lifespan: Early Cretaceous, 109.0 to 105.3 MYA. Location: Sofia River, Boriziny District (Port-Bergé), Sofia, Madagascar. Size: 5.5 cm.

*Origin of Species* in 1859, the more forward-looking workers adopted the ideal of a classification based on phylogenetic relationships."

All Mesozoic ammonites share an evolutionary history. Yet there are several superfamilies special enough to merit recognition.

Two sides of a *Phylloceras* ammonite preservation with pyrite clusters.

***Phylloceras* Suess 1865**
Order: Phylloceratida. Appearance and lifespan: Early Jurassic, 183 to 182 MYA. Location: Cingoli, Macerata Province, Marche, Italy. Size: 10 cm.

## Phylloceratina

**First Recorded Appearance: Early Triassic, ±251 MYA**
**Last Recorded Appearance: Late Cretaceous, ±70.6 MYA**

Instead of debating this creature's phylogenic placement, let us appreciate the superfamily Phylloceratoidea. These ammonites had worldwide distribution, and fossil formations indicate they clearly thrived in the long-gone Tethys Sea. Its appearance was streamlined; family members are appreciated for their smooth, involute, laterally flattened shells. Likely adorned with sinuous striations across the nacre, they encased a phragmocone supported by sturdy septa that imprinted fern-leaf suture patterns where they attached to the conch. The largest species of the superfamily Phylloceratoidea had a diameter equal to the length of an iPad. This ammonite's hydrodynamic design gave it the ability to glide through the water, not unlike the way a Frisbee soars through the air.

## Lytoceratina

**First Recorded Appearance: Early Jurassic, ±201 MYA**
**Last Recorded Appearance: Late Cretaceous, ±85 MYA**

Ammonite fans may wax poetic about the graceful Capricorn whorl that identifies members of the Lytoceratidae family. These cephalopods are characterized by tubular shells that grew rapidly and sometimes matured with space between each whorl. The conch's umbilical area was narrow and deep, nestled amid high suboval whorls

***Lytoceras* Suess, 1865**
Order: Lytoceratida. Appearance and lifespan: Jurassic, 164.7 to 161.2 MYA. Location: Atsimo-Andrefana, Madagascar. Size: 5.5 cm.

with rounded flanks. Surface ornamentation was subtle; flanks were typically striated with subtle, densely packed growth lines. Below the outermost nacre, sutures were complex and branching, and the animal had a living chamber that consisted of the better part of the last spiral. Partial to deeper water, Lytoceratoidea had almost oceanwide distribution for more than 100 million years, from the Early Jurassic to the early Late Cretaceous.

## Ancyloceratina

**First Recorded Appearance: Middle Jurassic, ±170 MYA**
**Last Recorded Appearance: Paleocene, ±61.6 MYA**

The most idiosyncratic-looking ammonites are likely the members of the suborder Ancyloceratina. Familiarly known as heteromorphs, as its name implies, this group changes as it matures. Juveniles had a variety of coiling types that aged into something that may have included flowing shafts, bends, and / or hooks. Rustic ribbing, rows of protuberances, and rose-stem spines were conventional ornamentation for these creatures. Sutures branched like aerial views of river deltas.

*Acrioceras* Hyatt, 1900
Order: Ancyloceratida. Appearance and lifespan: Early Cretaceous—132.9 to 129.4 MYA. Location: Sengily, Wolga, Russia. Size: largest is 7 cm.

Reef preservations such as this are reminiscent of Australia's Gogo Formation.

biosphere of early ammonoids, the reef preserved an array of fish and primitive sea scorpions known as eurypterids.

## We Dig Ammonoids: Fitzroy Crossing, Australia

The well-preserved fish of the Gogo Formation have taken precedence over ammonoids, but we'll venture about 100 kilometers (60 miles) north to Fitzroy Crossing. Nestled in this Aboriginal land is a cache of Frasnian faunal stage nautiloids and goniatites.

Communicate with the locals and let them know your intentions. They will share with you what is hiding in the rocks and where terrain is sacred land. Expect to find a worldwide marker specimen—*Manticoceras*, an internationally recognized genus of ammonoid included in the order Ceratitida. Available for discovery is a really

## Go to the Gogo

There's a special thrill when you find communities of fossils from specific orders. During the Devonian, oceanic biota teemed around rich reef complexes, which were known to ring the supercontinents. Earth has preserved an extraordinary example of a Late Devonian reef in the Kimberley Basin of northwest Australia. This incredible piece of world history captures a vivid tableau of a time from 382.7 to 372.2 million years ago, when Australia was still part of Gondwana. An early tropical equatorial ecosystem, the continental shelves were built on a framework of corals and sponge-like Stromatoporoidea at the time Ammonoidea were developing their efficient coil design.

Now a dry barrier reef located in flat, parched tropical grassland used for cattle ranching, the formation once extended more than 1,000 kilometers (621 miles) around the ancient supercontinent. The Gogo sediments reveal that, for a time in the Late Devonian, the seafloor was a hypoxic—low oxygen—environment. Beyond an extraordinary

Located in the Kimberley Basin of northwestern Australia, the Gogo Formation is an extraordinary example of a Late Devonian period reef conservation.

Image credit: Виктор В, CC BY-SA 3.0, via Wikimedia Commons.

# The Legalities of Fossil Prospecting in Australia

Australia has the 1986 Protection of Moveable Cultural Heritage Act to safeguard its treasures. At its essence, the terms of the act declare that there is no prospecting until you get authorization from the property owner or relevant authority. If you are in Australia to collect fossils, get your location releases in writing and be prepared to share this permission slip with the customs officials.

*Manticoceras* **Hyatt, 1884** (sp)
Order: Agoniatitida. Appearance and lifespan: Devonian, 382 to 372 MYA. Location: Atlas Mountains, Errachidia Province, Drâa-Tafilalet Region, Morocco. Size: 11 cm.

amazing collection of curiously sutured clymeniid ammonoids such as *Acanthoclymenia*, *Probeloceras*, and *Prochorites.*

As inviting as it may sound, Gogo Station is not something you listen to when you want to dance the Watusi or do the Loco-Motion. Down Under, in Australia, "station" is the word for "ranch." To get to this Late Devonian paradise, fly into Broome and take National Highway 1 east for 410 kilometers (255 miles). When you arrive, you'll find a protected formation of siltstone, shale, and calcarenite with numerous weather-resistant limestone concretions. Inside the treasure nodules may be one of a variety of ammonoids or other marine fauna. More than fifty species of well-preserved fish have been recovered; sometimes they are preserved in three dimensions due to rapid encasement and the slow rate of decay in oxygen-poor conditions.

Curiously, area fossils are prepared by dissolving the matrix using an acid solution. On rare occasion, the process exposes soft-tissue preservations, including an occasional placoderm fish embryo with an umbilical cord. Thanks mostly to the exceptional quality of preservation, scientists have been able to conclude that Devonian sea creatures gave birth to live young.

***Crioceratites loryi*** **(Sarkar, 1955)**
Order: Ammonitida. Appearance and lifespan: Early Cretaceous, 136.4 to 130.0 MYA.
Location: Drôme, Rhône-Alpes, France. Size: largest is 7 cm.

# Named for a God

Ammonoidea's worldwide distribution has made these marine cephalopods conversation pieces in both erudite circles and unexpected mainstream outlets. Check out the surroundings—mounted fossil shells are often in the background on the sets of popular TV series. On financial channels, they're sitting on office shelves while power brokers are interviewed. Ammonites may be part of the lobby décor at elegant hotels. And they grace many a coffee table in elegant homes. Fossilized cephalopods imply a level of style and sophistication in upper crust locations throughout the world, and apparently, it's always been that way. This cosmopolitan spiral-shelled predator was given a Greek name of an Egyptian god by a Roman professor. Millennia later, a Teutonic paleontologist named the subclass, while an Ivy League American and a Soviet scientist expanded the orders, and that's just the Western world.

## Deified by the Egyptians

Legends from ancient times associate ammonite fossils to the coiled rams' horns and the Egyptian sun god Amun. Divine and all-powerful, he was often depicted as a ram, with a ram, a man with the head of a ram, or a guy with ram's horns.

Amun emerged as one of the eight primordial deities of central Egypt during its Fifth Dynasty (around 2400 BCE). He was designated as the local divinity for the city of Thebes (where Luxor is currently located).

Sculptures of Amun at the Karnak Temple at Luxor, Egypt.

Image credit: Steve F-E-Cameron (Merlin-UK), CC BY-SA 3.0, via Wikimedia Commons.

The Kingdom of Ammon occupied the north central region of the Trans-Jordanian Plateau from the latter part of the second millennium BCE to at least the second century CE.

Image credit: Kingdoms_of_Israel_and_Judah_map_830.svg;* Oldtidens_Israel_&_Judea.svg; FinnWikiNo derivative work; Richardprins (talk) derivative work: Richardprins, CC BY-SA 3.0, via Wikimedia Commons.

In what seemed like the time it took for a papyrus plant to sprout, Amun was elevated to one of the most important gods of the Egyptian empire. He was considered Lord of All (who encompassed every aspect of creation). For more than a millennium, worship for Amun grew in strength and reach. The people of the Nile linked him with the god Ra, who reigned over the sun and was said to have created the world. Like atoms forming molecules, two deities became one and Amun-Ra was inaugurated as an all-encompassing deity who oversaw virtually every aspect of creation—both visible and invisible.

## Amun to Ammon

Trading expeditions shared the legend of Amun-Ra throughout the Mediterranean. In the Middle East, the country that is now Jordan then called itself the Kingdom of Ammon. From the latter part of the second millennium BCE to at least the second century CE, this ancient Semitic-speaking kingdom referred to its people as the "Children of Ammon" or "Ammonites." Today, Jordan's modern mecca of

Amman is located on the site of the ancient capital of Rabbath Ammon. Throughout Ancient Greece, Amun-Ra morphed into the ram-God, Zeus-Ammon. He was honored on money at the colony of Cyrene (now Libya) with a coin featuring the head of a god with curling ram's horns.

But all good reigns must come to an end. There was a revolt that brought Cleopatra to power and Amun-Ra was discredited. To add insult to history, in 27 BCE, an earthquake destroyed a number of the temples honoring Amun, facilitating the rise of the cult of Isis and the construction of the Greco-Roman world.

**Metatissotia Hyatt, 1903**
Order: Ammonitida. Appearance and lifespan: Late Cretaceous, 89.3 to 85.8 MYA. Location: Gebel Yelleg, Themed Formation, North Sinai Governorate, Egypt. Size: 18 cm.

**Crioceratites duvalii (Léveillé, 1837)**
Order: Ammonitida. Appearance and lifespan: Early Cretaceous, 136.4 to 125.45 MYA. Location: Nouterivren Nyons, Drôme, Auvergne-Rhône-Alpes, France. Size: 8.5 cm.

**Ammonite Bite:** The chemical ammonia gets in name from sal ammoniac, a crystalline salt that was derived from the dung of camels and used in Egyptian rites and rituals.

The word "ammonite" may evoke an Egyptian god, but Greek was used as the underlying language of this fossil's identification. Equally curious, the cephalopod subclass was given its Hellenic name by a Roman scientist. In 77 CE, Plinius the Elder released *Naturalis Historiae*, a prototype encyclopedia. The compilation is said to have described ammonites as "Hammonis cornu"—horns of Ammon (Amun). Memorialized in writing, these ancient spiral preservations spent the better part of two millennia being called the equivalent of Ammon's stone in a variety of languages.

Digging deeper, many genera of ammonites have names ending in *ceras*—think *Hildoceras, Dactylioceras, Australiceras*. The designations arose from the Greek word *keras*—meaning "horn." Ironically, Egypt's discoveries of this fossil subclass have been limited to the elusive Cretaceous cephalopod specimen, *Libyoceras*.

## We Dig Golden Ammonoids: Kielce, Poland

In south-central Poland, nestled by the Świętokrzyskie Mountains, is a massive quarry with a small

Highlighted in red on this map of Poland, Świętokrzyskie Voivodeship (province), is known as a hiding place for Paleozoic treasures.

Image credit: TUBS, CC BY-SA 3.0, via Wikimedia Commons.

section of thin-bedded marly limestone from the latter part of Devonian Period. Chockablock full of pyritized ammonoids and other treasures, the location yields some of the most spectacular golden goniatites found anywhere in the world.

Kudos to you if you can pronounce the name—Świętokrzyskie—it's something akin to *shin-tok-sin-ski*. Fortunately, the pronunciation may actually be more difficult than finding specimens in this ancient range. Better known to English speakers as Holy Cross Mountains, this fold belt is one of the oldest highlands in Europe. They were formed some 440 million years ago during the Caledonian orogeny. The late Devonian reef and basin system was part of a vast equatorial carbonate shelf off the southeastern coast of the primordial superconti-nent of Laurasia. Today, the Holy Cross range hides

a trove of fossil riches. The pocket we're interested in is the Kowala Quarry, and it's about the size of a 30-meter thick (100-foot) football field that yields goniatites from about 364.7 to 360.7 million years ago. Treasure specimens may be found along the forested west edge of the Morawica Mine, which is adjacent to a village of the same name. Here is where one finds *Discoclymenia cucullata*, with its sleek involute disc shape and a suture pattern that looked like an arrow pointing along the broad exterior of the shell. *Alpinites zigzag* had a thicker conch with a small, closed umbilicus at its mid-dle, and sutures that appear to be lobes and dips somewhat gone awry. *Posttornoceras contiguum* was wider, almost roundish, with simpler sutures and a keyhole design on the spine of the shell. Also to be found are pyritized Bactritida, brachiopods,

*Polonoceras planum* **Dybczynski, 1913**
Order: Goniatitida
**Orthoceras Bruguière, 1789**
Order: Orthocerida. Appearance and lifespan: Devonian, 376.1 to 364.7 MYA. Location: Kowala, Kielce County, Świętokrzyskie Voivodeship, Poland. Size: Goniatitida: 1 cm, Orthocerida: 2 cm.

gastropods, bivalves, fish fragments, tiny crustaceans, and small round nodules.

"In Poland, you can find Devonian ammonoids in many places, but pyritized specimens are discovered only in one, Kowala Quarry, near Kielce," shared engineer, collector, and field paleontologist Andrzej Gorsky, a specialist on this formation. "It's a giant quarry, (Permian, Carboniferous, and Devonian layers), but the stratum of pyritized ammonoids is small. Fossils are numerous, but mostly only fragments. Complete specimens are rare. The largest one I found was 4 centimeters [1.6 inches]. Mostly they are smaller than 1 centimeter [0.4 inch]."

Rich in natural resources and cultural history, the county of Kielce has been inhabited since at least the fifth century BCE. The area economy has been bolstered by the region's natural riches. Beyond fossils, it is a mining zone for limestone, copper, lead, and iron. Heavily utilized by civilizations for more than nine hundred years as a smelting center, Kielce has become a go-to location for those interested in the business and science of metallurgy.

## The Legalities of Fossil Prospecting in Poland

"In Poland, the topic of fossils is not fully regulated. There is no ban on searching for fossils, but there is an old law that says all fossils should be reported to the appropriate authorities, but it is a dead law that no one follows, and no one enforces," observed Andrzej Gorsky, a field paleontologist who regularly digs at locations around Poland. "The new law was passed several years ago. I can sell fossils, also abroad, whose unit value does not exceed 4,000 euros."

"When it comes to permission, of course, I have to obtain the owner's consent. If I enter the quarry without permission, I will be asked to leave without any consequences. But if I look for fossils in the forest, I may be fined heavily. The same applies to searching for fossils in areas where scientific research is carried out.

"As for the export of fossils from Poland. No problem with shipping via mail. I haven't heard about any problems at the border either. But if you encounter an uneducated customs officer, things may be different."

**Clymenia (Münster, 1834)**
Order: Clymeniida. Appearance and lifespan: Late Devonian, 376 to 359 MYA. Location: Kowala, Kielce County, Świętokrzyskie Voivodeship, Poland. Size: 3.5 cm.

## UNESCO

Kind of like your second cousin becoming president of the United States, the ammonite, or at least its appellation, is indirectly involved with the United Nations' inauguration of their World Heritage Program. The United Nations Educational, Scientific and Cultural Organization UNESCO seeks to build peace through international cooperation to preserve the planet's treasures. Their mission of saving places began in Egypt, where we again encounter the divine and all-powerful namesake of Ammonoidea. Built by Pharaoh Ramesses II, the temple complex of Ramesses, beloved by Amun at Abu Simbel was completed in 1244 BCE. It was one of several grand holy places constructed to show the kingdom's strength and assert its ideals over the Nubian people. More relevant to our story, it was one of the last remaining shrines to the once omnipotent Amun.

The temple's idyllic location posed a challenge to modern Egypt's growth and development.

The High Atlas Mountains are the tallest of a series of ranges that separate the Mediterranean and Atlantic coastlines from the Sahara Desert.

Image credit: Michal Osmenda from Brussels, Belgium, CC BY-SA 2.0, via Wikimedia Commons.

Leaders wanted to construct the Aswan High Dam to harness the power of the River Nile. The proposed reservoir would submerge the temple complex in the proposed Lake Nassar reservoir. It would have been an international outrage to have knowingly allowed the waters to destroy a shrine that predates Western religion. Thus, in 1959, an international campaign was launched to save the monuments of Egypt from being flooded by the dam construction. The UNESCO Nubian Salvage Campaign raised the necessary $40 million for the four-year relocation of the Temple of Ramesses, beloved by Amun. In 1964, a multinational team of archeologists, engineers, and heavy equipment operators successfully relocated the complex to a nearby desert plateau.

The experience at Abu Simbel emphasized that conserving valuable pieces of history needed to be a global effort. So, in 1972, the United Nations gave UNESCO the mission to protect the planet's cultural and natural heritage—and thus the World Heritage Site recognition program was born. Like Machu Picchu, the Eiffel Tower, and the Grand Canyon, UNESCO World Heritage Sites are globally appreciated for their outstanding universal cultural and/or natural value to humankind. These elite locations are preserved in cooperation with the public and private sectors. To date, the list of World Heritage Sites has grown to well over 1,000 locations involving around 170 countries. The designation has some powerful benefits. The title automatically increases tourism, and the area becomes eligible to receive funds, consulting, and protection against destruction during a war. We will visit several ammonite-related UNESCO sites throughout this book.

## Morocco Is Exceptional

If we go back to the time when the basement rock of Africa was formed—about 540 million years ago—the planet's landmasses were entirely different from what we know today. During the Cambrian Period, Australia, South America, India, Africa, and Antarctica were joined in the Southern Hemisphere, and collectively known as the *continent* of Gondwana. It was the dominant supercontinent at a time when oceans covered more than 85 percent

*Platyclymenia* ammonoid death assemblage.

*Platyclymenia* **Hyatt, 1884**
Order: Clymeniida. Appearance and lifespan: Devonian, 364.7 to 360.7 MYA. Location: Oued Chebbi East, Drâa-Tafilalet, Morocco. Size: largest is 7 cm.

of the globe (compared to 71 percent today). From the rare preservations that capture that epoch—Qingjiang, China, and Burgess Shale, Canada—we have learned that the waters were inhabited by tiny multicellular creepy crawlies—early examples of complex life. In similar Late Cambrian marine setting was *Plectronocerida*, a preliminary order of cephalopod with a ventral siphuncle running through the conch's internal compartments. This feature would give most shelled cephalopods the ability to control their buoyancy—a distinguishing characteristic of Ammonoidea.

As the ocean habitats evolved, hundreds if not thousands of miles of island continents crept toward the equator. Landmasses drifted north, geologically uniting to form Laurasia. This continent building event, known as the Caledonian orogeny, was a Paleozoic event that pushed up an epicontinental sea shelf and created what was to become Morocco's Anti-Atlas mountain range.

## We Dig Ammonoids: The Tafilalt Platform, Morocco

**Location: Eastern Anti-Atlas Mountains Morocco**
**Stratigraphic Range: Late Silurian—Devonian—Early Carboniferous**
**UNESCO Biosphere Reserve**
**Date of UNESCO Inscription: January 15, 2005**

Representing one of the world's finest tableaus of Devonian ammonoid evolution, the Tafilalt Plateau has laid the cephalopod symphony bare for all to treasure. Tucked into the southeastern part of Morocco's Anti-Atlas Mountains is

an extensive fossil record that spans six geologic periods, from 541 to 251.9 million years ago. Well weathered, worn down, and preserved like floors in a high-rise building, each level in this formation is its own exclusive biota capturing the evolution of Paleozoic ocean life. "It is always interesting to examine what happens around the origin of an important group," remarked Christian Klug. "My interest in cephalopods evolved because at first, I wanted to make a MSc [Master of Science] thesis on vertebrates, but then I got the exciting offer to do fieldwork in Morocco. So I wound up writing first my MSc and then my PhD about Early and Middle Devonian ammonoids. Morocco bears probably the richest record of early ammonoids, and hence it became an obvious choice for study."

This paleontological marvel of mesas and basins teems with preservations of ancient life. To the untrained eye, the bucolic Tafilalt landscape in the eastern Anti-Atlas Mountains is a view of palm groves stretching lazily along the River Ziz for as far as the eye can see. The largest Sahara Desert oasis in Morocco, it was an appreciated caravan respite along an arduous trade

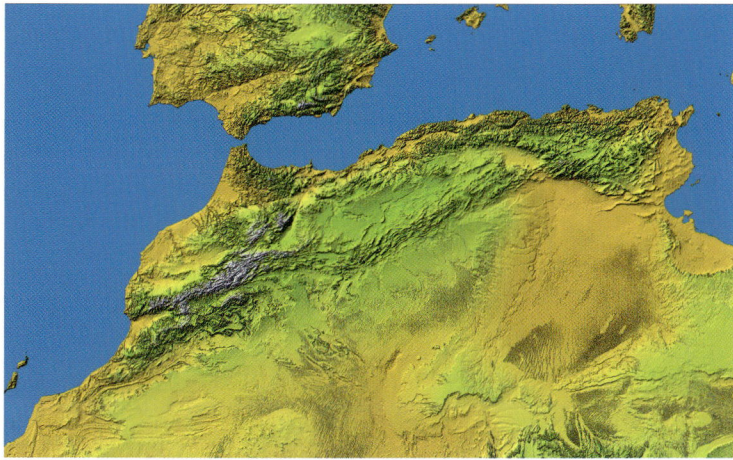

An aerial map of Spain, the Strait of Gibraltar, Morocco, and the Atlas Mountain ranges.

Image credit: Public domain, via Wikimedia Commons.

*Erbenoceras* was one of the earliest ammonoids.

**Erbenoceras Bogoslovsky, 1962**
Order: Agoniatitida. Appearance and lifespan: Devonian, 409, to 392 MYA. Location: Anti-Atlas Mountains, Drâa-Tafilalet Region, Morocco. Size: largest is 9 cm.

route for more than a millennium. Formerly fortified towns like Erfoud, Rissani, and Jorf allow one to envision the stories from the *Arabian Nights.* But what marvels paleontologists are the two expansive outcrops of Paleozoic rock massifs that have risen from the ocean floor. The Tafilalt captures extraordinary stone aquariums of sea life from the Cambrian to the Early Carboniferous periods.

Today, the area is subdivided into a mesa and two bowls. Running longitudinally across the region, the Tafilalt Platform segregated the ecosystems of the Maïder and Tafilalt basins. This fascinating story of evolution is told in formations that cover about 20,000 square kilometers (7,722 square miles). These layers of aquatic history capture orthocones, nautiloids, and ammonoids in such quantities that they occur in nearly rock-forming densities. The fossil-laden Tafilalt Platform is a Konzentrat-Lagerstätte because of the abundant concentration of fossils on-site.

"The Devonian oceans must have been an amazing place—a warm, planet-spanning marine world filled with ammonoids, menacing predators like jawed fish, and thousands of spinose trilobite species which had developed their elaborate external ornamentation as a defense mechanism to ward off those ever-more-prevalent undersea threats," observed Andy Secher, field associate at the American Museum of Natural History and author of *Travels with Trilobites.*

Likely the most cephalopod-rich Konzentrat-Lagerstätte preservation discovered to date, the Tafilalt Platform captured a cornucopia of Ammonoidea genera that flourished and disappeared during the numerous Devonian transgressions. This preservation began about 430 million years ago, while the Anti-Atlas Mountains were still forming. During this Silurian once-upon-a-time, the seas had low oxygen conditions, and marine life struggled to survive. Ultimately, these early cephalopods—like the straight-shelled *Plagiostomoceras*—were entombed in a pinkish limestone seam that's about as thick as a guitar is tall. These pointy, tapered Orthocerida were the older cousin to the burgeoning ammonoid.

A Devonian seascape featuring straight-shelled **Orthocerida** and newly minted subclass **Nautiloidea.** Orders: Nautilida, Orthocerida. Appearance and lifespan: Late Devonian, 375 to 359 MYA. Location: Filon Douze, Errachidia Province, Drâa-Tafilalet, Morocco. Size: Nautilida, less than 1 cm;, Orthocerida, 4.5 cm.

**Ammonite Bite:** At the time Temperoceras ruled the oceans, sea creatures basically had two modes of life. There were pelagic faunas that populated the upper tiers of the sea, usually by the continental shelves. The other forms of life were the benthic animals that survived on the sea bottom.

Partial to deep subtidal marine shelves, orthocone cephalopods remained a constant part of the Tafilalt seascape until this marine environment was no more. These early creatures fossilized along with sea lily-like crinoids and the occasional bivalve.

Looking above that stratum, you can actually see the point where shelled cephalopods appeared. The star of this stage was *Temperoceras*, a straight-tested, slow-moving predator that soon made up about 25 percent of life in the region's shallow seas. True to conched cephalopods, the test was reinforced by a series of internal chambers. The additional support allowed these early nautiloids to control their buoyancy, making them some of the earliest scuba divers. The luxury of being able to manipulate their position in the oceans gave *Temperoceras* environmental domination. They thrived; some genera grew so large, they, reached beyond the height of modern-day human beings. So prevalent was this group that a horizon of the Tafilalt is named in its honor. "I've been buying, selling and trading Moroccan fossils for more than fifty years, and I'm still excited when I see a specimen captured in time," shared commercial paleontologist Bill Barker, founder of the Sahara Sea Collection. "It's mind boggling that some of these creatures are close to a half billion years old."

## Ammonoid Horizons

World events regularly caused ammonoids and other creatures to suffer low oxygen conditions and slowly suffocate, leading to episodes of significant marine extinction. Preserved as mass mortality ribbons, these layers capture a specific point in history. Understanding the age of each band of ammonoids allows scientists to date the surrounding fossils more precisely. These geologic strata are often named for the predominant ammonoid taxa found in the matrix. For example, the *Pinacites jugleri* layer was designated for the narrow-keeled, involute agoniatite that predominated from 392 to 390 million years ago. All other creatures found in that preservation band date to that time frame. A bit higher up in history's layer cake is a noteworthy section bed called the *Sellanarcestes* layer. It is remarkable because it captures a marine episode known as the Daleje Transgression, which snuffed out the Agoniatitida order, embalming the last of the group in a matrix of slate clays and lime marls. Rise to another level, and the formation reveals when the order Clymeniida arose, how it proliferated and when it expired. The bull's-eye whorled *Platyclymenia annulata* dominated the matrix some 370.6 to 364.7 million years ago. "The annulata zone undoubtedly represents a peak in the evolution of Paleozoic ammonoids," noted cephalopod specialist Dr. Dieter Korn, curator of the Museum für Naturkunde Berlin.

When you arrive at the Tafilalt, what's old is new, and fossils are everywhere. Four significant areas of deposits developed in the Anti-Atlas: the Maïder Platform, the Maïder Basin, the Tafilalt Platform, and the Tafilalt Basin. The distinctive fossil concentrations assist scientists in weaving together the story of our planet's past. For example, the two basins of Paleozoic rocks reveal the vestiges of the union between the landforms of Laurussia and Gondwana. Its fossiliferous layers hold keys to the planet's natural history.

## Basin Logic

Because it is shaped like a big bowl, it's easy to extrapolate why the Late Devonian Maïder Basin Konservat-Lagerstätte was the ideal breeding ground for several orders of Ammonoidea. In the calm waters Clymeniids thrived: there is the endemic *Muessenbiaergia sublaevis*, with its sublime spiral, and the armadillo-shell-patterned goniatite known

This *Temperoceras* limestone fossil tabletop, excavated at Morocco's Tafilalt Plateau, captures a transgression that occurred some 400 million years ago. The transition line shows anoxic black matrix on one side and a better oxygenated brown milieu on the other. Notice that coiling cephalopods are only in the lighter matrix. Tabletop size: 122 cm.

as *Imitoceras*—a species that continually radiated, died, and fossilized in the carbonates, marls, and claystone sediments. The basin's deposits from the end of the Devonian have yielded abundant nonmineralized cephalopod jaws of several Paleozoic genera. These local discoveries confirm that the entire *Ammonoidea* subclass likely had some type of substantial mouth apparatus. "All ammonoids had jaws," confirmed Klug. "The presence of these beaks/jaws in several lineages in the Late Devonian represent good evidence for this."

The Maïder discoveries have also allowed paleontologists to distinguish organic soft-body parts inside ammonoid body chambers—an amazing accomplishment because many specimens are smaller than a bottlecap. Beyond ammonoids, orthocerids, and nautiloids, the treasures of this ancient lagoon are accompanied by some carbonized algae plants and articulated arthropod skeleton remains. The abundance of soft body tissue

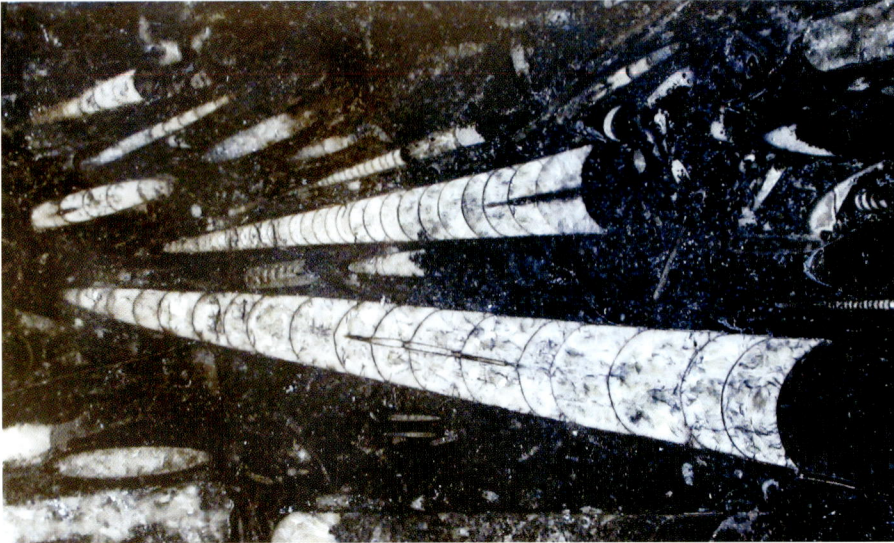

In the High Atlas Mountains of Morocco, regional economic prosperity benefits from the sale of fossil-laden, Devonian black limestone, such as this polished orthocerid block.

*Orthoceratidae* **McCoy 1844**
Order: Orthocerida. Appearance and lifespan: Early Devonian, 409.1 to 402.5 MYA. Location: Filon Douze, Errachidia Province, Drâa-Tafilalet Region, Morocco. Size: largest is 14 cm.

is why the southern Maïder has been awarded the status of Late Devonian Konservat-Lagerstätte.

The distance between the Maïder and the Tafilalt basins was less than 50 kilometers (30 miles) as the crow flies, yet the ecological differences grew increasingly more evident as the Devonian progressed. Not only did the lagoons diverge in sediment thickness and facies, but each location also developed proliferations of distinct species of cephalopods during the same time stage, which is rather extraordinary if you stop to contemplate the concept of environmental adaptation. The dynamic Devonian palaeoecological differences in the two bowls as well as their preservations reflect the special position and paleoenvironment of the Maïder Basin.

## Black Death

The transgressions and regressions that spotted the Devonian Period were just the first season of the central Pangean mountain-making series. Plate tectonics rattled the late Paleozoic Era, acting up some 375 million years ago and causing clatter between Laurussia and Gondwana that lingered on for some 40 million years. Finally, they joined and together they formed Pangaea. Today,

remnants of this ancient mountain range include the Little Atlas Mountains of Morocco and the Appalachian Mountains of North America and the Scottish Highlands.

During this continent-building episode, sea life was constantly adapting to shifting conditions. Along the Tafilalt, the struggle to survive low-oxygen sequences is apparent in the thick successions of jet-black limestone thatched with white cephalopods. It's a commercially quarried layer in the southeastern Maïder region, above the *Platyclymenia annulata* Zone. It dates back 372 million years ago, when everything here died from lack of oxygen. The decomposition of the high volume of organic matter is what created the deep dark color so prized by collectors. Truly, if you can disassociate yourself from the uncontrollable yet distressing circumstances of the late Paleozoic extinction, these Moroccan fossils in matrix transform into rather grand pieces of organic art. Fossil sculptures and finished goods have made this stratum a source of regional economic importance. Known as the *Temperoceras* Black Limestone, this stratigraphic interval also yields remains of the jawed placoderm fish *Dunkleosteus*. In the Maïder, complete skeletons have been found in nodules as large as a car.

# The Legalities of Fossil Prospecting in Morocco

"The US recently repatriated several vertebrate fossils to Morocco," noted Bill Barker, who has made a career exporting Moroccan fossils. "Invertebrates have never been a thing."

Moroccan laws do allow artisanal mining by locals, and more than 50,000 citizens do so. Fossil prospecting, trade, and export brings in more than $40 million annually. To you, the visiting paleontologist, that means hire a guide and get a receipt.

International trade of any fossil specimens is technically prohibited by Moroccan law, and export of "objects of anthological or archaeological interest" are illegal. However, the laws are often insufficient or not applied in practice. "Show you bought stuff and contributed to the economy. The authorities like that," concluded Barker.

Do appreciate the variation of suture patterns among these Paleozoic ammonoids. Included in this grouping are:

***Cheiloceras subpartitum* (Munster, 1839)**
Order: Goniatitida. Appearance and lifespan: Late Devonian 376 to 359 MYA. Location: Kowala, Kielce County, Świętokrzyskie Voivodeship, Poland. Size: 2 cm.

***Latanarcestes* (Schindewolf, 1933)**
Order: Agoniatitida. Appearance and lifespan: Early Devonian, 408 to 393. MYA Location: Errachidia Province, Drâa-Tafilalet, Morocco. Size: 2.5 cm.

***Discoclymenia atlantea* (Korn et alii 2016)**
Order: Goniatitida. Appearance and lifespan: Late Devonian, 372 to 365 MYA. Location: Fezzou, Drâa-Tafilalet, Morocco. Size: 3 cm.

***Cymaclymenia* Hyatt, 1884**
Order: Clymeniida. Appearance and lifespan: Late Devonian, 364.7 to 360.7 MYA. Location: Errachidia Province, Drâa-Tafilalet Region, Morocco. Size: 3 cm.

***Discoclymenia cucullata* (von Buch, 1832)**
Order: Goniatitida. Appearance and lifespan: Late Devonian, 364.7 to 360.7 MYA. Location: Jebel Ouaoufilal East, Errachidia Province, Drâa-Tafilalet Region, Morocco. Size: 3 cm.

# The Great Dying

The late Paleozoic series of extinction events are collectively referred to as the Great Dying. It is rumored to be Earth's most severe set of elimination events to date. Creating the margin between the Paleozoic and the Mesozoic eras, as well as the Permian and Triassic geologic periods, were several distinct pulses that triggered the disappearance of up to 96 percent of all marine species and 70 percent of terrestrial vertebrate varieties. "In the oceans, the combination of the increase in temperature, loss of oxygen, and acidification contributed to the loss of life we see at the end of the Permian," noted Pedro Monarrez, a paleobiologist at the University of California, Los Angeles.

Noteworthy for ammonoid transitions was the Kellwasser Event, also known as the Frasnian-Famennian Extinction. It has been speculated that the chaos may have been sparked by underwater volcanism, the growth of terrestrial plants, and/or global climate cooling. Another theory posits that it was partly caused by the Siljan meteorite impact in the area of present-day Sweden about 376.8 million years ago. Whatever the causes, the outcome was a global shift of elements throughout the oceans. Evidence in Australia, Europe, China, and Africa show an abrupt swing in carbon, sulfur, and oxygen isotopes, as well as a sudden loss of organic matter.

"The F-F [Frasnian-Famennian Stage] Boundary Event consisted of two intervals: the Lower and Upper Kellwasser Events named after black limestone horizons in Germany," explained John Catalani of the Mid-America Paleontology Society (MAPS). "During the Lower Event, sea levels continued to rise from the Taghanic Onlap, but that was followed by a drastic fall in sea levels. Similarly, the Upper Event saw an initial rise in sea levels followed by an even more drastic fall

**Ammonite Bite:** The Paleozoic Transcontinental Arch refers to the islands of a burgeoning North America. It extended from New Mexico to Minnesota and through the Great Lakes region.

that offlapped the Transcontinental Arch and exposed carbonate platforms, thus decimating those ecosystems. The F-F Event was somewhat protracted over 1 to 2 million years and displayed a stepwise extinction pattern. Along with the continued collapse of the coral / stromatoporoid reefs, brachiopods and ammonoids were severely affected."

The fallout from the Kellwasser Event saw the gradual extinction of the extensive goniatite suborder Gephuroceratina, which includes the beloceratids and manticoceratid groups. Also disappearing were many conodont species, most colonial corals, several groups of trilobites, and various brachiopod genera.

The lack of oxygen in the water resulted in lower diversity, but a phoenix rose out of the ashes of the Kellwasser Event. It sparked the ascent of a new and stronger ammonoid model. "Only one group, the Clymeniida, has a high rate of phylogenetic diversification during the Famennian," noted Korn in the research paper "Ammonoid evolution in Late Famennian and Early Tournaisian."

**Ammonite Bite:** It was after the Kellwasser Event when ammonoids dropped down the food chain and became prey. Cephalopods of this stage have been found with teeth-like bits from surviving species of eely conodont jawed fishes.

*Platyclymenia intracostata* **Frech, 1897**
Order: Clymeniida. Appearance and lifespan: Devonian, 375 to 359 MYA. Location: Kowala, Kielce County, Świętokrzyskie Voivodeship, Poland. Size: 3 cm.

## The Clymeniida Connects

Truly one of the missing links uniting nautiloids and ammonoids, Clymeniida was created with a hybrid siphuncle. Similar to the more advanced cephalopods of 370 million years ago, this cephalopod had camarae walls that were convex, positioned toward the living chamber. Also like the others, Clymeniida's buoyancy control apparatus was initially positioned along the interior wall of the whorl. After the first few septa, however, the calcareous tube that enclosed the siphuncle migrated to a dorsal position on the inside of the coils, opposite that of most ammonoids and akin to more primitive shelled cephalopods. "Both Famennian groups, clymeniids and goniatites, are distinguished by the position of the siphuncle [which played a role in locomotion]: it is located at the ventral side of the whorl in goniatites and at the dorsal side of the whorl in clymeniids," wrote Korn.

Clymeniida disappeared in the wake of a global disturbance known as the Hangenberg Black Shale Event. Another incident of the Great Dying, it took place about 358.9 years ago, when there was rapid sea-level fall due to glaciation in the Southern Hemisphere. The redistribution of water lead to the accumulation of vast amounts of organic carbon in continental shelf sediments. Unable to weather the changes, several groups of Goniatitidae, all the clymeniid ammonoids, and *Phacopid* trilobites disappeared, among others. Their fossils have been preserved in the dark sediments, which is also laden with orthocones, small trace fossils, bivalves, and remnants from other sea life of the stage. In total, about 57 percent of all biological families and 83 percent of all genera disappeared as a result of deadly

The Salt Range is part of an active fold and thrust belt, formed in response to the collision between the Indian and Eurasian plates.

Image credit: Nadeem Ali, CC BY-SA 4.0, via Wikimedia Commons.

conditions. Trilobites, which helped to define Paleozoic seas, didn't make it through the transition.

An ascending step on the evolutionary ladder, Clymeniida was on this Earth for a relatively short time—around 10 million years. They were limited in range and have been discovered only in Europe, western Asia, North Africa, and possibly Australia. Various genera were indigenous to a single defined basin or other circumscribed zone, such as an inland sea. Clymeniida's life span was likely limited because of the endemic proliferation of species and the ongoing oceanic crises.

In the aftermath of this Paleozoic annihilation, shelled cephalopods and the rest of the battered families strived to recover. Vertebrates were slighter than before, but the upside was that these smaller taxa were able to successfully diversify, filling the niches left in the extinction's wake.

## We Visit Ammonoids: The Salt Range and Khewra Salt Mine

**Mianwali Formation**

**Location: Punjab, Pakistan**

**Age: Permian-Triassic**

**Date of UNESCO Submission: December 2016**

An exotic place to see how goniatites radiated ceratites is along the Salt Range in Pakistan. Plate tectonics raised mighty geomorphic mountains containing hidden primordial ocean treasures. The activity began around the start of the Mesozoic Era. Tectonic activity between the Indian and Eurasian plates raised the open sea environment. What was once ocean floor became land, a mesa crowned by a fold-and-thrust belt of mountains. Now known as the Pothohar Plateau, areas rise more than 1,500 meters (4,992 feet) above sea level. A fine example of paleo stratigraphy, many layers are clearly exposed, allowing excellent opportunities to appreciate the metamorphic features of this highland, which sits just south of the Himalayas.

Embedded between the bright red Precambrian marls of the Punjab province are some of the world's most abundant seams of rock salt, as well as coal, gypsum, and other desired minerals. Mineral veins are interspersed with ocean preservations rich in ammonoids and other early marine fauna. This substantial sedimentary succession has sections from a variety of periods—Precambrian to Cambrian and Carboniferous to Quaternary, with access to fossiliferous Permian and Triassic areas. With its well-stratified rocks preserving the secrets of several geologic stages, the Salt Range captures the transition from Paleozoic to Mesozoic sea life.

**Ammonite Bite:** Although the Salt Range spans east to west, sedimentary strata uniformly slope to the north.

Map of India highlighting the Punjab region.

Image credit: TUBS, CC BY-SA 3.0, via Wikimedia Commons.

(right) **Agathiceras sundaicum Haniel, 1915** (sp)
Order: Goniatitida. Appearance and lifespan: Permian, 295.5 to
279.3 MYA. Location: Hatu Dame, Aileu District, East Timor.
Size: 2 cm.

## Goniatites Radiate Ceratites in Paradise

In the wake of the events surrounding the Great Dying, *Goniatitida* were all but eliminated. Merely a few genera from the order Prolecanitida transitioned through the crisis. In recovering seas, Ceratitida progressed into the power order. More complex than its predecessors, this state-of-the-art model was more resistant to hydrostatic pressure. The repetitive undulating suture patterns looked a bit like a line of jigsaw puzzle piece tabs. The evolved complexity of these arch-and-trough seams strengthened the shell and may have facilitated buoyancy. Another noteworthy update was in the jaw: Ceratida refined the ammonoid's signature aptychus-like jaw, allowing for more diverse feeding.

Oceanic predators, these cephalopods readily propagated and filled a variety of nektonic marine environments. As an order that survived some 100 million years, ceratitids were at their most prolific in the Late Triassic, when there were more than 150 genera. This period was their time as the radiant order. By the end of the period, there were fewer than ten groups remaining. The history of Ceratitida can be found throughout Pakistan's 300-kilometer (186-mile) long linear chain of sheer escarpments, jagged peaks, rolling hills, desolate ravines, and fertile valleys, collectively known as the Salt Range. The formation captures the transition between Permian and Triassic fauna, a history of what happened in the wake of the Great Dying.

**Neoicoceratidae Hyatt 1900**
Order: Goniatitida. Appearance and lifespan: Permian, 314.6 to 272.5 MYA. Location: Aidaralash Creek, Asselian, Kazakhstan. Size: 3 cm.

## We Dig Ammonoids: The Mianwali Formation, Pakistan

The tectonic deformation of the Salt Range has given us a highly fossiliferous record of marine sediments spanning the Paleozoic–Mesozoic transition. Break down the deepwater geochronology of the Mianwali Formation, and you will find seven unique layers. On the rebound from extinction events, Permian species tend to be scarcer; however, there are still outcrops with aggregations of obscure Prolecanitida species and holdover orthocerids. Truly, the reason to prospect the Mianwali is to capture the progress of ceratites. Several strata have ammonoid species in their descriptive names. Stratigraphically important genera include *Prohungarites*, which was involute with a sharp keel. *Tirolites* had evolute spirals with patterned pleats and nodules, similar to many of the Mianwali genera. *Prionites* sloped toward the keel, while *Nammalites* was compressed on its sides. The lineage of these ceratites was incestuous, with most genera descending from the superfamilies *Prolecanitoidea* and *Meekocerataceae*.

Within these well-defined paleostratigraphic bands, ammonoid adventurers will be awed by the rich accumulations of genera like *Glyptophiceras*, which gracefully widened its ribbed evolute whorls as it grew. *Anasibirites* was an involute genus with delicate striations. *Wasatchites* was involute with struts wrapped around the ammonoid's face toward the venter, which gave texture to its round shape. *Gyronites* was streamlined; flat and involute, it moved easily through the water. Change locations and another shale stratum will have discoveries like *Arctoceras, Anakashmirites, Meekoceras, Stephanites, Owenites, Anasibrites,* and *Pseudaspidites*, all of which died in close proximity to one another. In addition to previously known taxa, the Salt Range has, to date, introduced five new genera—*Kyoktites, Ghazalaites, Pashtunites, Awanites* and *Subacerites*—with eighteen new species. Optimistically, there is still more to be discovered. Other Pakistani Salt Range areas like Nammal Nala (Early Triassic), Chiddru (Permian), and Amb (Permian) are more obscure locations for digging well-preserved ammonoids.

*Waagenina subinterrupta* **Krotow, 1885**
Order: Goniatitida. Appearance and lifespan: Permian, 290.1 to 279.3 MYA. Location: Aktasty River, Aktobe, Aktobe Region, Kazakhstan. Size: 3.5 cm.

Rohtas Fort, near Jhelum, Pakistan is a UNESCO World Heritage Site.

Image credit: Hussain Khalid, CC BY-SA 4.0, via Wikimedia Commons.

## The Legalities of Fossil Prospecting in Pakistan

Hypothetically, certain items are forbidden and are controlled for export from Pakistan; antiques (not antiquities) are on that list. Yet, in 2021, Pakistan exported $4.85 million in antiques, making it the twenty-eighth largest exporter of heirlooms in the world. Neither fossils nor minerals are mentioned in the country's 2022 Export Policy Order or in the Antiquities Act of 1975.

If you are carrying your fossils home with you, customs controls exports, and it can be an arbitrary decision from one agent to the next. Have some kind of paperwork that shows you paid for something related to your rocks and contributed to the local economy. This may also be a good time to shed extra rupees.

If you are looking for less cosmopolitan intrigue, the attorneys at Josh and Mak International stealthily hint, "Senders are responsible for making sure the destination country accepts the goods they are shipping."

*Owenites* **Hyatt & Smith, 1905**
Order: Ceratitida. Appearance and lifespan: Early Triassic, 251 to 247 MYA. Location: Crittenden Springs, Elko County, Nevada. Size: largest is 3 cm.

## Khewra Salt Mine

While you're in the neighborhood, visit the nearby Khewra Salt Mine, also known as the Mayo Salt Mine. Legend has it that this massive formation was identified sometime around 327 BCE, when Alexander the Great (king of an ancient Greek realm) explored the Jhelum and Mianwali region on his Indian campaign. The vein was discovered when the army's horses began licking the rock salt stones. Eureka! The site has been precious ever since. During the Mughal Empire, the salt was traded in markets as far away as Central Asia. Upon the downfall of that realm, the profitable business was taken over by the Sikhs. The mine continues to operate to this day. We know the quarry for its pink sodium chloride (NaCl), often marketed as Himalayan salt. The excavation is a major tourist attraction, drawing up to 250,000 visitors a year.

## Jhelum

To get to the ceratite sites on the Salt Range, visit the closest city, Jhelum, which is located an hour-and-a half from Islamabad, Pakistan, and a three-hour drive from the heart of the Punjab—Lahore, India. Once you are in Jhelum, it's around 100 kilometers (60 miles) to the Salt Range. While you're in town, appreciate how the old city is an ancient maze of narrow streets and traditional bazaars. If you follow where they lead, more likely than not you'll wind up at someplace interesting. The heart of the Land of Martyrs and Warriors, Jhelum is upstream from the site of the 326 BCE Battle of the Hydaspes. About 18 kilometers (11 miles) to the northwest is the sixteenth-century fort, Qila Rohtas. Now a UNESCO site, it was built by an Afghan king to protect his holdings after he took control of the Mughal Empire. More recently, this

***Discoceratites dorsoplanus* Spath (1934)**
Order: Ceratitida. Appearance and lifespan: Middle Triassic, 242 to 237 MYA. Location: Muschelkalk, Schwäbisch Hall, Stuttgart, Baden-Württemberg, Germany. Size: 15 cm.

disciplined metropolis of about 200,000 is lauded for providing soldiers to the British Indian Army prior to Pakistan's independence. Today, it is home to the Jhelum Military College and headquarters for the country's armed forces. The region has a monsoon-influenced, humid subtropical climate—it is extremely hot and humid in summer, and cold and dryish in winter.

**Ammonite Bite:** Near the Salt Range is Tilla Jogian, a Hindu pilgrimage center established in the first century BCE. Members of the disciplined order at this site are noted for their ear piercings.

***Phylloceras* Suess, 1865**
Order: Phylloceratida. Appearance and lifespan: Early Jurassic, 182 to 175.6 MYA. Location: Yorkshire, England, United Kingdom. Size: 15 cm.

# Ammonites Are Tools to Mark Time

n his 1586 book *Britannia*, cartographer William Camden observed, "If you break them you find within stony serpents, wreathed up in circles, but eternally without heads." Remarkably, paleontology and the concept that land was once under water did not take hold in Europe until the Age of Enlightenment. It was nearly two hundred years after Camden's observation when French paleontologist Georges Cuvier evaluated and contrasted living animals to fossilized creatures, establishing the fields of comparative anatomy and paleontology. His study made its way across the English Channel, where it became a source of conversation and debate among erudite British men of science.

The English-speaking world can look to the dapper gentlemen of the Geological Society of London as serving as the rock-solid foundation for paleontological thought. Those early nineteenth-century British scholars made a truly historic discovery while exploring the striated exposures that mark the cliffs along the Yorkshire coast. As their inquiring eyes ascended the area's extensive Mesozoic outcrops, they recognized something extraordinary. The ammonoid species embedded in the site's defined sedimentary layers changed every few feet, and wherever they searched along that coast, the cephalopod strains appeared in the exact same chronological order. The cliffs held the saga of ammonite evolution.

Papers were published about this astonishing discovery. Word of these geologic breakthroughs slowly found their way around the globe.

Ammonite stratification revelations were shared via state-of-the-art nineteenth-century mediums—books were published, wires were sent, reports were mailed. In time, a cosmopolitan coterie of paleontologists came to realize that ammonite stratification along Yorkshire's coast was also reflected within their local rock formations. The evidence of the layered cephalopod evolution defined along England's now-famed Jurassic Coast was found to hold true in other parts of the planet—whether that was in Germany, Argentina, China, or the United States. About as fast as anyone can properly pronounce *Reynesocoeloceras praeincertum*, the science of biostratigraphy was born. This new academic endeavor served a vital role in organizing the international study of ammonites and allowed scholars to date the age of any layer in a formation with accuracy and assurance.

"Ammonites make the best index fossils. They evolve so rapidly, are so abundant, and are so widespread that once you pick up an ammonite, you know exactly where you are in the stratigraphic section," explained Neil Landman, curator emeritus of fossil invertebrates, Division of Paleontology, Richard Gilder Graduate School of the American Museum of Natural History in New York City. "In some areas, like Montana and South Dakota, the Upper Cretaceous is subdivided into something like sixty ammonite zones, giving you a precise location in time. Combined with radiometric dates from interleaved ash beds, you can

***Scaphites whitfieldi* Cobban 1951**
Order: Ammonitida. Appearance and lifespan: Cretaceous, 94.3 to 89.3 MYA. Location: Fall River County, South Dakota, United States. Size: 3.5 cm.

immediately pin your location down to an absolute age expressed in millions of years. (Of course, thanks to the years of work of many geologists.)"

The chronology of shelled cephalopods is akin to a grand staircase with a different type of *Ammonoidea* on every step. The conchs of these animals evolved and vanished so quickly within the fossil record that certain species lived only during a specific paleontologic stage. Thus, an ammonite has the unique ability to delineate a precise time in geologic history. When pieced into a context with other local marine fauna, scientists are able to solve the puzzle of the surrounding rock's geologic age.

"Ammonites can be used to distinguish intervals of geological time of less than 200,000 years'

**Ammonite Bite:** Malacology is the branch of invertebrate zoology that focuses on the study of molluscs. Teuthology is the branch of marine zoology that concentrates on living cephalopods. Ammonite specialists have been known to refer to themselves as ammonitologists.

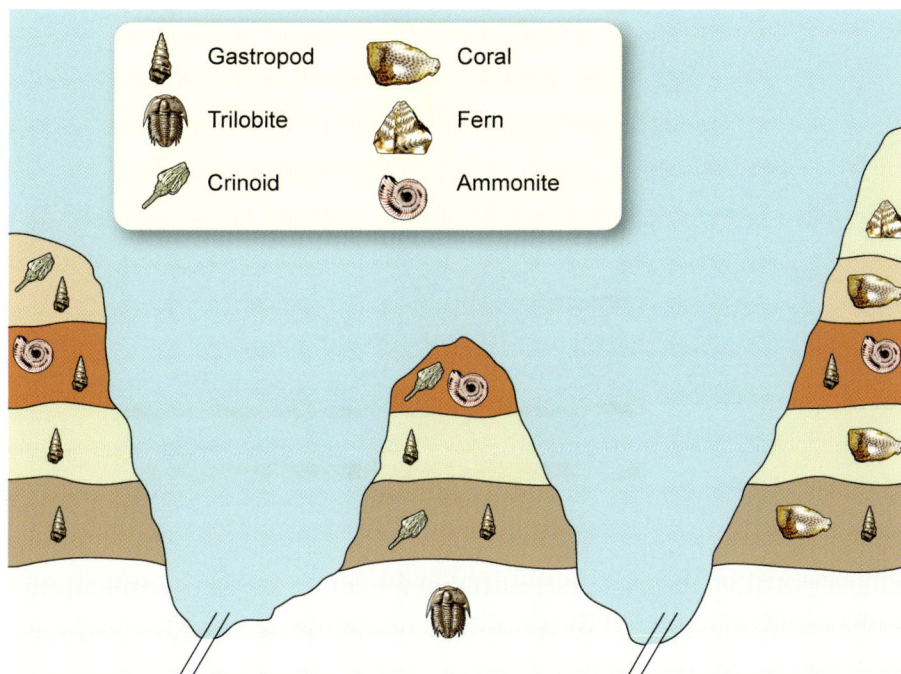

Stratification occurs in most sedimentary aggregations, as well as igneous rocks formed at Earth's surface, typically as the result of tectonic activity and volcanic occurrences.

Image credit: Kurt Rosenkrantz, CC BY-SA 3.0, via Wikimedia Commons.

duration. In terms of Earth history, this is very precise," observed a spokesperson for the British Geological Survey. "The rapidity of ammonite evolution is the single most important reason for their superiority over other fossils for the purposes of correlation. Such correlation can be on a worldwide scale."

Many members of the ammonite family Scaphites resemble nature's version of the number 9. The distinctly chambered uncoiled shell makes members of this genus an excellent Cretaceous-era index fossil and gives this curious-looking cephalopod a distinguished place in natural history as well as in collector's cabinets.

To date the layer where an index ammonite is found, paleontologists must dig past that genus to the cephalopod species. For example, if *Tropites subbullatus* is encountered, geologists know that the stone in that zone is between 221.5 and 232 million years old. Conversely, if the prevalent ammonite is *Trachyceras aonoides*, the surrounding matrix is from 232 to 235

**Ammonite Bite:** Index fossils are specimens that are so singular in appearance that they can confirm a specific point in the fossil record and hence an age.

million years old. And if *Eoprotrachyceras curionii* is found, the adjacent milieu is between 235 and 242 million years old. We know this thanks to biostratigraphy. When certain indicator species are found in key sedimentary outcrops, they serve to confirm the geologic period of the surrounding deposits.

## Index Fossils

Designation as an index fossil is a unique honor in the paleontological universe. In many ways, it is the fossiliferous equivalent of winning the Super Bowl, and you go in history books as having been the most distinguishable and widespread species

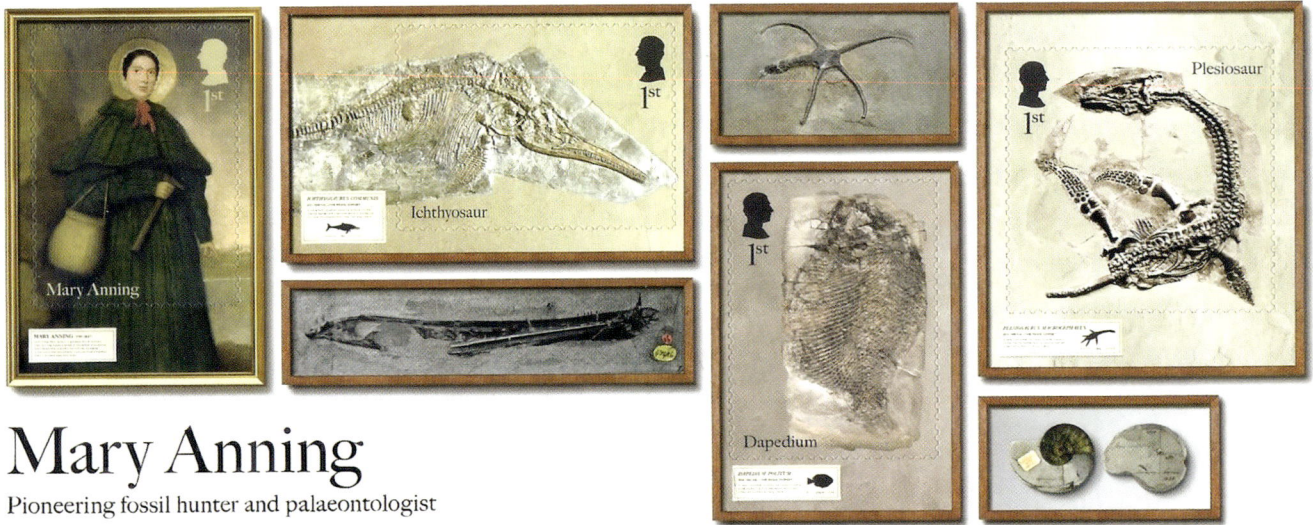

"Mary Anning: The Age of the Dinosaurs" stamp set, released by the British Royal Mail, 2024.

# Mary Anning
Pioneering fossil hunter and palaeontologist

of your time. Such a distinction guarantees the selected species an esteemed position in the natural history hall of fame. Each of these important indicator species share several essential attributes:

1. Reasonably common with wide geographic distribution.
2. A paleontologically short duration on Earth.
3. Rapid evolutionary trends.
4. A relatively recognizable species.

In contrast to other invertebrates, *Ammonoidea* were sensitive to the slightest environmental changes—orders, genera, and species were forced to regularly adapt or die out due to shifting marine conditions. Like the age rings on a tree, this cephalopod's rapid ecological assimilation reveals a chronological history of geologic events. Each layer divulges the details of a distinct stage of our planet's evolution. Analyzing the particulars of these timeless tales has allowed researchers to link various ammonoid species to sea-level changes, extinctions, migrations, and radiations.

## Mary Anning

Renowned British paleontologist Mary Anning has been a woman of inspiration, conversation, and curiosity for centuries. Urban legend has it that the Victorian tongue twister, "She sells seashells on the seashore," was about the woman who wandered the beaches of Lyme Regis. That tidbit of information is about as factual as the 2020 romantic biopic *Ammonite*, which implied that the Georgian-era fossil collector was sexually fluid. It is all conjecture. This self-taught "princess of paleontology" was an unlikely candidate to change history, yet she did.

A prolific English fossil hunter and amateur anatomist, Anning is credited with the discovery of several marine reptiles that advanced the fledgling field of paleontology. Her early nineteenth-century findings came at a time when most people in England still believed in the biblical account of creation, which inferred that Earth and the creatures that inhabited it were just a few thousand years old. The fossils she uncovered raised questions about the history of living things and of the planet itself.

Should one have visited her village along the English Channel during the late Georgian era, they'd likely have spotted Anning on the eastern part of the beach, looking for fossils around the base of Church Cliff. A humble, unassuming woman, she'd comb the shores of Lyme Regis dressed in a long coat with a heavy wool wrap for added warmth and a bonnet on her head. She would have had a wicker basket in the crook of her arm, a geologist's hammer in hand. Most likely, you would not have realized that she was one of the finest paleontologists of her day.

Born into a working-class family, Anning grew up in a house built on the town bridge, a shanty that would flood during nasty storms. Her father was a carpenter and cabinetmaker, but to supplement the family income, he and his children, Mary and Joseph, collected specimens from the fossil beds along the local beaches. The Annings combed the coastline, gathering ammonites and other small treasures to sell to tourists as souvenirs. By the time Mary was six years old, she was a regular presence by her father's side along the shore. She knew the most favorable time for fossil hunting was after a storm. The foul weather usually caused bits of cliff to fall to the beach below. The concretions would often break open on impact, revealing ammonites and other prizes she could sell to tourists.

The Annings were outcasts of sorts. Mary's parents were religious dissenters—Protestants who separated from the country's requisite religion, the Church of England. The national faith found it of no importance to educate the lower classes. Conversely, the Congregationalist church the Annings attended emphasized schooling for the poor; this allowed the children to attend ecclesiastic Sunday school. Mary was taught to read and write the holy doctrines. She then used those skills to learn all she could about her paleontological discoveries.

After their father died of tuberculosis in 1810, brother Joseph followed in his trade as an apprentice upholsterer. Mary continued hunting for ocean treasures. Near the coach stop at a local inn, she set up a table of curios and curiosities and sold fossils to visitors, collectors, and scholars. These souvenirs were given intriguing local names. Ammonoids were called snake stones or Ammon's horn, tapered belemnites were known as devil's fingers, and spinal vertebrae were nicknamed verteberries. The first time Anning sold a large ammonite concretion to a woman for half a crown (around 25 U.S. dollars in today's currency), it confirmed paleontology as her future. Anning got to know the cliffs in a manner that few people had before, and she uncovered a sea monster of sorts: the first complete ichthyosaur fossil ever found, and it rocked the British Empire. On exhibition in the capital, the aquatic reptile generated tremendous interest and raised creation questions among the country's population and press. The "Crocodile in a Fossil State" was later sold for a bit more than £45 to Charles Konig of the British Museum. He named Anning's discovery *Ichthyosaurus*. (The skull of the specimen is still at the Natural History Museum in London.)

Credited with unearthing several orders of Mesozoic marine reptiles that changed our understanding of how the world evolved, Mary Anning's finds proved to be an important step in advancing humankind's understanding of fossils in relation to geology and biology. Yet Anning certainly didn't fit the Georgian model of a fossilist. In the early nineteenth century, it was a man's England; women were not yet allowed to vote, hold public office, or study at university, and they were certainly not allowed to attend the meetings—as members or guests—of the increasingly influential Geological Society of London. In the 1800s, the only occupations open to working-class women were farm labor, domestic service, or toiling in newly opened and not particularly safe factories. It was rare for a female to become a "proper" scientist, yet Anning's amazing discoveries allowed her into this elite group of wealthy and educated males. Today, her legacy is commemorated at the Natural History Museum in London and in her hometown at the Lyme Regis Museum, where there is a gallery dedicated to her life. In a pleasing coincidence, the museum stands on the site of her birthplace and family home.

"Mary Anning is a very important part of fossil hunting history in Charmouth," observed resident Lizzie Hingley, founder of Stonebarrow Fossils. "When I was young, (and there are many of the other collectors on the coast here who are older than me), Mary was not as renowned as she is today. Now, she is a great inspiration and it's fantastic that she is now so famous! In the UK, her story is now a part of school curricula. If time travel were possible and Mary came to 2023, she would be amazed about the new fossil preparation techniques but likely saddened by the lack of potential on the beaches. There are so many people looking now that virtually nothing gets destroyed by the sea. Should a fossil hunter from today travel back

The English Channel separates the United Kingdom from Continental Europe. At its narrowest point through the Strait of Dover, the channel is approximately 21 miles (34 kilometers) from shore to shore.

Image credit: User:Bastin (original author) and User:Antoby, CC BY-SA 3.0, via Wikimedia Commons.

to see Mary in her time, they would be amazed at the wealth of preservations it was possible to find. But they'd be restricted by the tools available to collect and prepare fossil specimens."

## The West Dorset Fossil Collecting Code

In some parts of the world, fossils are considered the property of the state. In others, locations regulations may make it difficult for amateur collectors to participate. This is not true in West Dorset. All paleohunters are needed along the Jurassic Coast. In this unique location, fossils are revealed on the overnight and may be destroyed by the sea later that same day. Residents, paleontologists, and property owners are unable to save even a fraction of these treasures, but they want to try. Along the Heritage Coast, the public is requested to help gather as many fossils as possible. The area has adopted a progressive collectors' code. It applies to

preserved specimens gathered between the mean low water mark to the back of the beach. "The most important thing is that we save the fossils," emphasize the spokespeople at the Charmouth Heritage Coast Centre.

Fossil hunting is all about being in the right place at the right time. The longer you spend searching, the better your chance of finding something interesting. Be prepared to preserve your newfound treasures. You'll want to carry a backpack with some small bags, plastic wrap or aluminum foil, and containers to keep the fossils you find intact. When splitting concretions, wear eye protection. The local experts suggest you go with a guide who knows the key discovery locations along the Jurassic Coast. Contact the Charmouth Heritage Coast Centre or Lyme Regis Museum, as both regularly conduct guided fossil hunting walks. You can also hire a guide through the chamber of commerce.

## The Legalities of Fossil Prospecting in Southern England

The West Dorset Fossil Collecting Code of Conduct applies to the seventeen-kilometer stretch of coast between Lyme Regis and the village of Burton Bradstock, plus East Devon. It was devised by various stakeholders—a consortium of scientists, landowners, government agencies, collectors, and cultural and environmental organizations. It clarifies ownership and ensures that key specimens are preserved for posterity. "Category I fossils include new species or those specimens which may represent new species, fossils which are extremely rare such as the Charmouth dinosaur *Scelidosaurus* and fossils that exhibit exceptional preservation," according to the code.

Should one discover something extraordinary and wish to sell their distinctive specimens, they must first offer it to museums. Once the scientific institutions indicate that they don't have the budget and pass on the fossil, it can be sold to private buyers. Be advised, details of the sale must be reported. Contact the Charmouth Heritage Coast Centre for additional details.

The majestic Jurassic Coast Cliffs at West Bay, Chesil Beach, capture falling sea levels from some 175 million years ago.

Image credit: Robert Powell, CC BY-SA 3.0, via Wikimedia Commons.

*Caenisites tumeri* (also known as *Euasteroceras* **Donovan 1953**) *Promicroceras* **Spath, 1925** (sp) *Cymbites* **Neumayr 1878** (sp) Order: Ammonitida. Appearance and lifespan: Early Jurassic, 196.5 to 189.6 MYA. Location: Popcorn Stone, Charmouth, United Kingdom. Size: largest is 4 cm.

## We Collect Ammonites: The Jurassic Coast

**Location: Southern England, United Kingdom**
**Age: Triassic, Jurassic and Cretaceous**
**Date of UNESCO Inscription: 2001**

The prolific nature of England's Jurassic Coast fossils is indelibly intertwined with the tale of local paleontologist Mary Anning. She was the first individual to document the abundant fossil treasures hidden along this cherished 150 kilometer (95 miles) stretch of southern coastline. One of the busiest shipping lanes in the world, this northeastern extension of the Atlantic Ocean separates England from mainland Europe. A dynamic oceanscape, at 560 kilometers (350 miles) long, it is the smallest of the shallow seas around the continental shelf of the European Union.

Of particular interest to ammonoid collectors is the south coast, where exposures provide a continuous sequence of Triassic, Jurassic, and Cretaceous rock formations, spanning some 185 million years of Earth's history. The bluffs on this part of the coast are constantly eroded by the English

Channel seas, causing sections to crumble away. As a result, it is Europe's largest coastal landslide area, regularly experiencing cliff falls and mudslides and always revealing new fossils in the process. With each high tide, rock is worn off the steep ridge, exposing clues to the creatures that once swam in this ancient ocean. The constantly changing opportunities make the Jurassic Coast coves extraordinary fossil-collecting locations.

Curiously, during the Triassic Period, this area was a desert. And in the Cretaceous, it was covered by swamps. The key time frame here is the Jurassic, when the British landmass—then located around the same latitude of Morocco—lay submerged under a tropical sea. The sultry marine environment bore a rich marine ecosystem laden with life from ammonoids to massive reptiles. As geologic time progressed, the sea receded, and the Eurasian tectonic plate floated north. Around Lyme Regis and Charmouth, paleostratigraphers have been able to use ammonites and their rapid evolution of species to identify seventy-one layers of rock delineating 185 million years of history.

Expect to find familiar British genera like *Harpoceras*, *Eleganticeras*, *Hildoceras*, and *Asteroceras*, among so many others.

Erosion of the chalk cliffs has created features like horseshoe-shaped Lulworth Cove, the limestone arch known as the Durdle Door, and various carved stacks and pinnacles that now highlight nature's activity of the Jurassic Coast. The sea has spared the highest point on the south coast of Britain; sitting between Bridport and Charmouth is Golden Cap at a noteworthy 191 meters (627 feet). Some coastal defenses have been put in place in populated villages like Charmouth and Weymouth, but in other areas, the coastline remains in a natural state, thus allowing nature to take its course. While today's fossil hunters may imagine that they are following in Mary Anning's paleontological footsteps, the areas she combed have long since been washed into the sea.

"These places can be pretty dangerous, and fossil hunters should listen to the advice of locals, their tour guides, et cetera, in order to remain safe," observed Kate LoMedico Marriott, Earth and Environmental Sciences adjunct lecturer. "A friend of a friend was crushed to death by falling rocks here a few years ago."

Once you begin your search for ammonites, try not to get overwhelmed by the options. Many fossils simply lie exposed along the beach, like pebbles waiting to be spotted. No special gear is needed; you can find them using your eyes. Look for iron pyrite geodes, shiny round gray stones about the size of a grapefruit. Split the sphere with a rock hammer and, if you made the right choice, you may have one or more 190-million-year-old perfect spiral ammonite embedded within.

To be found here is the many-whorled *Psiloceras*—one of the earliest known Jurassic ammonites. This genus defines the base of the period. Another area favorite is the cosmopolitan *Dactylioceras*. One of the most prolific ammonite lineages, the presence of this evolute index ammonite with tubular coils identifies a preservation as Early Jurassic. Also in the mix is the attractive *Eleganticeras*, with its wavy shell pattern and sharp keel. These fossils may be discovered in a range of sizes, from tiny specimens the size of a thumbnail to those with diameters as vast as the space between your outstretched arms. This is the ideal locale to blaze your own trail and seek your own treasures. Anything on the beach is fair game, but it is forbidden to hack into the cliffs.

"I never start preparation on a fossil with an image of how it should be finished. I like to develop the end result as I uncover and understand the rock. Each fossil is unique, both in its preservation and position within a nodule, and as such should be prepped to bring out its highlights," shared Hingley, who has developed a distinctive presentation style for Jurassic Coast preservations. "I collect nodules whole, often looking for signs of ammonites on the outside. Many of them turn out to be empty, but I try to avoid using a hammer as it will create a break in the rock which will always be seen. I was self-taught in fossil preparation but have a degree in art and grew up in an artistic family. I like to think that this has influenced my style of preparation."

Lyme Regis is easy to reach. If you're going by train, the nearest railway station is at Axminster, five miles to the north; it connects to the Pearl of Dorset via the brightly painted Jurassic Coaster double-decker bus, which regularly runs up and down the seaside. If you are driving, take the A3052 roadway from Exeter.

**Ammonite Bite:** The Green Ammonite Beds of the Dorset Coast are named after *Androgynoceras lataecosta*, which has chambers filled with greenish calcite.

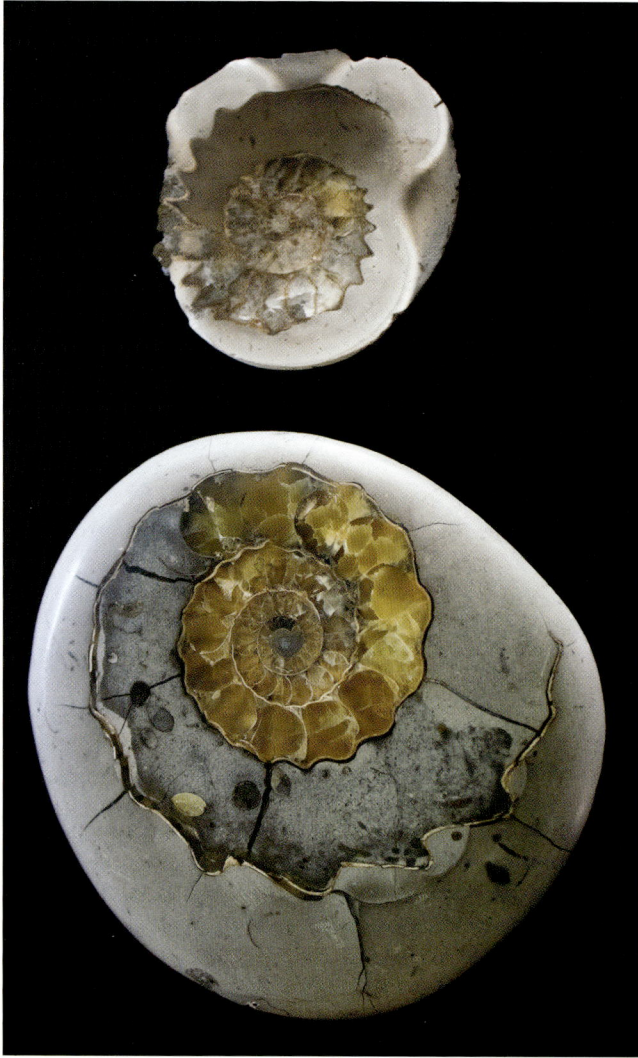

England's Dorset Coast is renowned for its green ammonite nodules.

**Liparoceras cheltiense (Murchison, 1834)**
Order: Ammonitida. Appearance and lifespan: Early Jurassic, 191 to 183 MYA. Location: Lower Lias, Blockley, Gloucestershire, England, United Kingdom. Size: 4.5 cm.

**Androgynoceras Hyatt, 1867**
Order: Ammonitida. Appearance and lifespan: Early Jurassic, 191 to 183 MYA. Location: West Dorset, England, United Kingdom. Size: 6.5 cm.

The damp winter months are when the finds tend to be most prolific. Wear a headlamp or carry a lantern because the sun goes down early. Exploring may be a slow, cold, and windy process, so dress to stay warm.

**Ammonite Bite:** When going onto the beach, always check the tide tables. You'll likely have the greatest success if you arrive at the beach as the ocean ebbs out. Plan to get safely back on dry land well before high tide.

The Jurassic Coast offers a remarkably rich geologic history and is one of the very few UNESCO World Heritage Sites where you can legally keep the ammonites you collect.

*Gymnotoceras blakei* (Gabb, 1864)
*Longobardites zsigmondyi* (Bockh, 1874)
*Tropigastrites lahontanus* Smith, 1914
*Tozerites gemmellaroi* (Arthaber, 1911)
Order: Ceratitida. Appearance and lifespan: Triassic, 247 to 242 MYA. Location: Faurett Formation, Pershing County, Nevada, United States. Size: largest is 3 cm.

## Discerning Order from Disorder

Biostratigraphy puts *Ammonoidea* in a chronological order that has been verified by scientific study, but confirming the names and lineage of these creatures can be quite another story. When it comes to ammonite details, experts reach different decisions. Modern ammonite specification is largely based on suture patterns. During the days of the Cold War, in the Soviet Union, Comrade Professor V. E. Ruzhencev introduced generic terminology for ammonoid suture line descriptions. He counted the undivided lobes and saddles characteristic of the Paleozoic cephalopods and gave each node a simple, universal name:

V = Ventral Lobe
L = Lateral Lobe

D = Dorsal Lobe
U = Umbilical Lobe
I = Internal Lateral Lobe
O = Omnilateral Lobe

As the paleontologist who used this strategy to name the order Agoniatitida, Ruzhencev explained, "In initial Agoniatites there is a ventral lobe and the dorsal one is absent (VO), whereas in primitive *Clymenia* there is a dorsal lobe and the ventral one is absent (UD)." This is logical for Paleozoic species, but when you describe Ammonitida using this method, it looks like an algebraic equation and is subject to error. The suture pattern terminology for the Jurassic superfamily *Stephanoceratoidea* is $(V1V1)UU^2II^1D$; Perisphinctoidea looks like $(V1V1)UU^1U^2II^1D$. Essentially, a tiny

A sunset view of Lake Shasta and its surrounding hills.

Image credit: Abshreve, CC BY-SA 3.0, via Wikimedia Commons.

bump or lack thereof separates superfamilies. Now consider that among the same species, various stages of development may exhibit different suture lines, or that endemic locations may have their own variations.

Cephalopod specialists at universities throughout the world regularly ponder the significance in the nobs and dips of ammonite sutures. Yet even research partners can be at odds about details, resulting in more than one name per species. To add to ammonoid identification confusion, scientists make a habit of reclassifying genus and species labels. Did you know the superfamily formerly known as *Perispinctaceae* is now called *Perisphinctoidea*? The change appears in *some* of the literature. Names can be different in different parts of the world. Technology only seems to add to the confusion. Science can now provide more information than previous cephalopod researchers ever had available. Additional details have led to the subdivision of orders, and a whole lot of trickle-down name-changing has occurred. But there's a basic rule of thumb, "Orders end in oidea, Suborders end in atina, Superfamilies in ataceae, Families in itidae," observed Cretaceous ammonite expert Neal L. Larson.

**Ammonite Bite:** The United States Geological Survey (USGS) has been the only scientific agency that functions within the Department of the Interior. Using the disciplines of biology, geography, geology, and hydrology, the USGS studies the country's environmental resources and the natural hazards that threaten the nation's well-being.

## We Dig Index Ammonites: Lake Shasta, California

Like miners searching for gold, the heroes of the USGS have combed the land of the United States and documented the country's natural assets for posterity. More successful than most gold diggers, their research uncovered bounty of a different kind.

The adventurers of the USGS map the paleontology and geology of our country. Among the things they found was the Late Triassic fauna of the Hosselkus Formation in Northern California. An obscure sedimentary seam in the global scheme of things, it peeks out of remote outcrops in Plumas

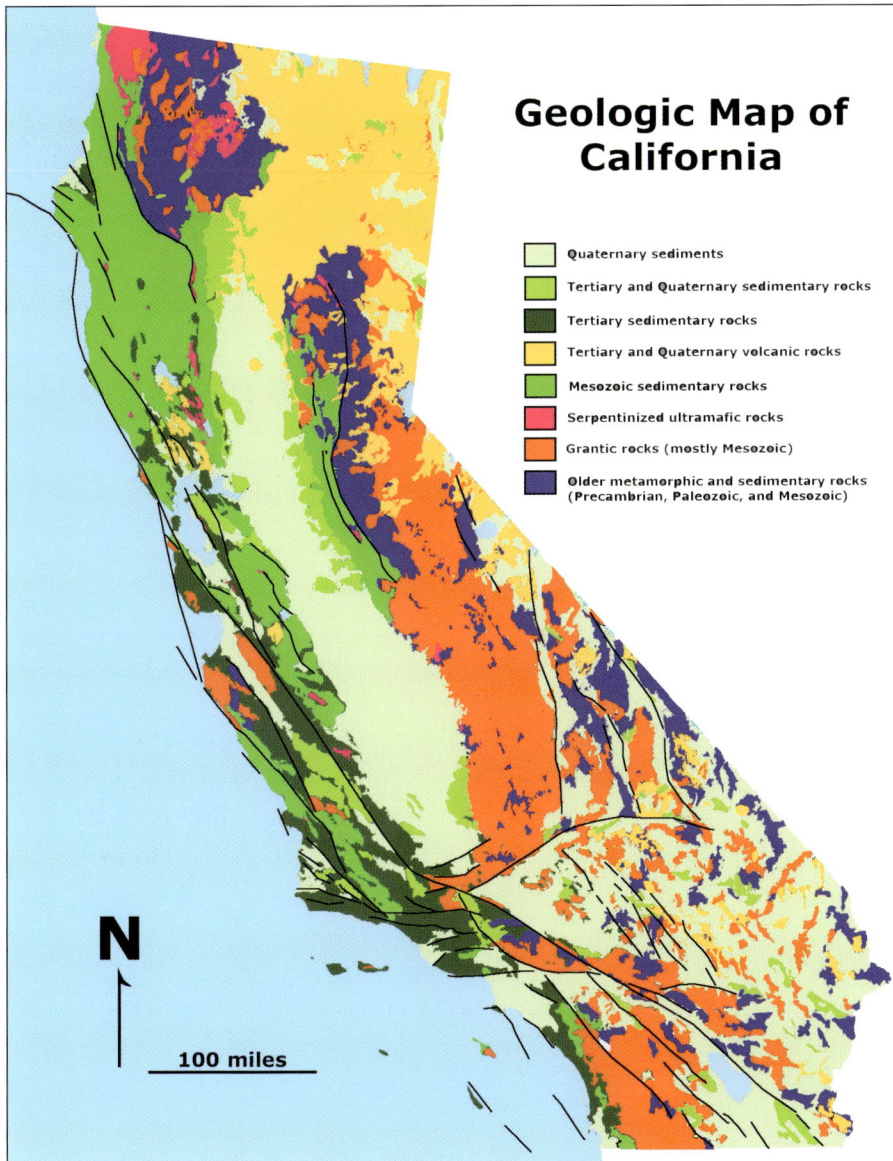

# Geologic Map of California

- Quaternary sediments
- Tertiary and Quaternary sedimentary rocks
- Tertiary sedimentary rocks
- Tertiary and Quaternary volcanic rocks
- Mesozoic sedimentary rocks
- Serpentinized ultramafic rocks
- Grantic rocks (mostly Mesozoic)
- Older metamorphic and sedimentary rocks (Precambrian, Paleozoic, and Mesozoic)

N

100 miles

Geologic map of California.

Image credit: USGS, public domain, via Wikimedia Commons.

and Shasta counties. These isolated repositories feature dangerous topography, the type that only attracts fools, thrill seekers, and science geeks. If you happen to fall into any of these three categories—or all of them—then the Hosselkus is an ideal locale for you.

This deposit is relatively narrow—about 43 meters (140 feet) thick—and ranges in age from 227 to 237 million years. What the locality doesn't possess in Mesozoic mass, it more than makes up for in its inherent paleontologic variety. To

date, well over two hundred species of invertebrates have been identified and described from the Hosselkus Limestone Formation and nearby Brock Mountain. The ammonite species here hail from the *Tropites subbullatus* zone, a weathered, jagged section of hillside that used to be Triassic ocean floor. *Tropites*—the indicator genus of the echelon—is recognizable by its spherical shell and central keel. Also lurking within the formation are Tropitoidea superfamily members—subglobular, involute samples of *Paratropites*, *Discotropites*,

The Hosselkus Formation outcrop is a stony anomaly in a bucolic landscape.

Image credit: David Henderson.

and *Paulotropites*, as well as *Juvaites*, *Sagenites*, *Leconteia*, *Trachyceras*, *Clionites*, *Arpadites*, and *Arcestes*. All are very tantalizing if you are looking for species within a particular vein.

For whatever reason, in paleontology circles, vertebrates inevitably remain more popular than invertebrates. Extrapolating on that supposition, most of the area's corresponding fossil search and rescue efforts have centered on the unique fish/lizard ichthyosaurs, *Shastasaurus* and *Californosaurus*. But if you're into ammonites, the Hosselkus Formation is nothing less than cephalopod paradise.

Retrieving Hosselkus specimens is full of tests and obstacles, the likes of which would challenge even Indiana Jones on a great crusade. Besides its remote location, these fossiliferous areas reside on National Forest Service land. Be aware that, prior to removing these primal preservations, you need permissions, permits, and all the accoutrements that go with collecting within an isolated area that is accessible only by a lake.

"Any work in the Hosselkus Research Natural Area (RNA) requires a permit from the Pacific Southwest (PSW) Research Station," explained Dennis Veich, emeritus geology program manager for Shasta-Trinity National Forest. "Even I need a permit to do anything up there and the RNA is in my forest. Despite the fact that it's a difficult area to reach, the RNA is a fossil hunter's paradise and well worth the effort to get there. As well as finding a metric crap-ton of ammonoids up there (still uncollected), I have found some ichthyosaur remains, another one of my favorite extinct critters."

Government lands, government paperwork: if you pay taxes, you are familiar with how it all works. The official permit application for collecting fossils at the Hosselkus Formation is the easy part. You must be clear about what you plan to collect. Do note that just writing "fossils" doesn't cut it. If you've received a permit to dig for ammonites, that does not grant permission to excavate any vertebrate material you may stumble over, no matter how appealing it may be. When you fill in the dates on your form, remember that you are venturing into the hinterlands. Allow yourself time to explore and appreciate the area's natural wonders properly.

As is typical with so much government paperwork, it's the attachments that require extensive time and effort. To dig fossils on managed land, you need a purpose as well as a permit. Each

*Discotropites* **Hyatt & Smith,1905** (sp)
Order: Ceratitida. Appearance and lifespan: Late Triassic, 232 to 221.5 MYA. Location: Hosselkus Formation, Lake Shasta, California, United States. Size: 5 cm.

beyond your backpack full of digging gear. During a recent expedition, it was our forest geologist who suggested expanding our team with local specialists and enthusiastic graduate students. He advised us of the best way to get to where we were going and how to get there. All we needed was to arrange for the boat.

Anyone who has visited California knows that, from its rugged mountains and redwood forests to its rocky coves and sprawling beaches, the state features some spectacular terrain. This majestic topographic variety is in part because the Golden State contains eleven distinct geomorphic provinces. These areas are where the North American Tectonic Plate has overcome any opposing force that dared challenge it. In these regions, the continental crust has been folded, deformed, and uplifted to build mountain ranges and ravines. Three of these unique geologic features happen to intersect at Lake Shasta, forming the network of canyons that is home to the Hosselkus Formation.

"The Klamath Mountains to the east are the great and ancient core of this geomorphic province. They provide a glimpse into the 'pre-California' past as they host terrains forged in remote regions of the ancient world," elaborated Professor Randal Reed, chair of the Earth Sciences Department at Shasta College. "It was time and tectonics that brought them to California, suturing them to the western North American plate as exotic features that stand out in contrast to the coastal mainland. Uplifted by subsequent tectonism and dissected by the evolving modern rivers of Northern California,

submission package must contain a letter from an accredited museum or academic institution stating the fossil fueled intentions of your expedition. If available include a paleontology curriculum vitae for each member of your crew, plus any maps and charts you'll need to clearly state your invertebrate intentions. Once the paperwork is out of the way, the next step is to figure out what else is needed

the Shasta Lake region now provides windows into this past across the ridges that separate its bays and inlets."

Some 38 million years ago, volcanoes began to erupt throughout the areas north and east of Lake Shasta. Now known as the Cascade Mountains, these lofty vents grew to become part of the American Cordillera. (FYI: This impressive sequence of mountain ranges forms a nearly continuous boulder-strewn backbone from northwestern North America to southwestern South America.) An arc in the legendary Pacific Ring of Fire, on occasion, the Cascades still sputter and surge. The activity is well documented at nearby Lassen Peak, which erupted in both 1914 and 1921. Just to the north is Mount Shasta, one of the most active vents in the entire Cascade Range. This conically shaped stratovolcano is second only Washington State's Mount St. Helens, which displayed its disruptive power in 1980. All this tectonic activity also plays into California's legendary reputation for earthquakes.

To the south and west of Lake Shasta, the Sierra Nevada Mountain Range developed around a granite core, where two of the planet's lithospheric plates smashed together for the better part of 150 million years. In the process, the terrain across these areas was abused, spewed, scorched, crumpled, rotated, and shuffled. Little of anything that was fossilized prior to this upheaval survived this reign of tectonic chaos.

While volcanic activity and plate tectonics transformed much of Southern Oregon and California, the area around Lake Shasta was largely spared because it lay between two developing mountain ranges. Northern California and Nevada

**Ammonite Bite:** The California Gold Rush occurred in the western foothills of the Sierra Nevada Mountains from 1848 through 1855. Because of its inaccessibility, the range was not fully explored until 1912.

***Clionitites* Strand, 1929** (sp)
Order: Ceratitida. Appearance and lifespan: Triassic, 235 to 216 MYA. Location: Hosselkus Formation, Lake Shasta, California, United States. Size: 5 cm.

***Trachyceras* Laube, 1869** (sp)
Order: Ceratitida. Appearance and lifespan: Triassic, 237 to 232 MYA. Location: Hosselkus Formation, Lake Shasta, California, United States. Size: 5 cm.

(where the accompanying outcrop is known as the Star Peak Limestone) represent the only areas in the continental United States where you can find this distinctive biozone.

As most paleontological prospectors have come to learn, fossil collecting tends to be easiest in the spring. Storm, snow, and ice may have cleaved and split a location's sedimentary rocks, revealing fossilized fauna and flora. If you plan to try your hand in the field, perhaps attempt early in the season, before the unpleasant summer nasties like rattlesnakes, ticks, and poison oak reawaken from their dormant state.

At the Hosselkus Formation, the Shasta-Trinity National Forest geologist proved to be invaluable in navigating the outcrops that yielded the Late Triassic ammonites of the *Tropites subbullatus* zone. The location was peaceful and pristine, and all it took was a few minutes of cleaving, cracking, and chiseling in the exposed rock before the ammonites we were seeking were uncovered. "The Hosselkus Formation is unique because the species of animals we find there have not been found anywhere else in the world," shared Veich.

The location is layered with swelling discs of involute *Discotropites*, *Sagenites*, and bulbous nautiloids. Specialists familiar with the area's unique preservation have calculated two individual subzones. The particular area we were exploring carries an abundance of *Trachyceras* ammonoids, among other fossil finds. Whatever one may discover—*Anatropites, Mierotropites* or any of the more than two dozen species of *Tropites*—it is an incredibly satisfying experience. The ground is awash in imprints of ancient ammonites that resemble suns in stone. Shells come out of the ground as a striped spiral with a decorative fringe. These are particularly pleasing finds because of their inherent aesthetic value as well as the minimal follow-up required for proper presentation. (Stabilizing the brittle limestone and cleaning up the rock is the prep work that follows discovery.)

## The Legalities of Fossil Prospecting on Federal Land

According to the US Bureau of Land Management (the original BLM), "Rules are generally more restrictive for wilderness areas, wilderness study areas, areas of critical environmental concern, national monuments and national landmarks, or historic and prehistoric sites and districts, so it's best to ask first." You are dealing with the federal government, so it is all about paperwork, permits, and permissions. Plus, you do not have authorization to sell items found on protected US lands.

**Ammonite Bite:** You know you've had a good day fossil collecting when you find so many fossils that you also discover what our geologist lovingly calls leaverite—meaning, "You can leave that piece of rock right there, we already have better."

An example of a leaverite fossil.

## What Have We Found?

The Hosselkus Limestone features a stunning variety of *Tropitidae* family members. Among the many genera, there's *Paratropites*, which was designed with a distinctively shallow umbilicus. There's the more hydrodynamic, saucer-shaped *Discotropites*; the knobby involute *Gymnotropites*; the smaller *Microtropites*; *Margaritropites* with its notable keel; and . . . you get the idea. Figure each genus has produced several local species, so two dozen or more different radiations are hidden in the rocks. As one attempts to identify recent fossils finds, they have the curious quandary of distinguishing which *Tropitidae* relative has been uncovered on this ammonite adventure.

Discerning what's what often falls beyond the purview of many collectors and most online ammonite groups. The "devil of" proper ammonite identification often relies on the details, not to mention the tabs and pockets of the suture pattern. The traditional technique for properly documenting fossils finds begins with noting the location of your dig: geotag the coordinates and photograph the section of the formation in which you'll be excavating. As you gather, collect a working group of specimens from the location, some for display and some for study, if possible.

As Sherlock Holmes said to his loyal sidekick after a day in the field collecting fossils, "It's sedimentary, my dear Watson."

When you have returned to the comfort of your prep area, do a basic cleaning and then sort your finds by their shape and size. Chances are fossils with similar conch geometry belong to the same genus. Sorting specimens is an artful business for several reasons. Some organisms died as juveniles; others were old. Creatures were either male or female. Not to mention there was natural variation, just like we find in *Homo sapiens*. You're likely to second-guess yourself as to whether you're looking at multiple examples of a single variable species or two different sorts of ammonites. Good luck counting those ventral and lateral lobes.

A microscopic view allows for the examination of suture patterns. Or you can digitally slice it in half and study whorl growth. By cross-referencing your information with USGS reports like *The Triassic Cephalopod Genera of America* and *The Upper Triassic Marine Invertebrate Faunas of North America*, the identity of the find becomes clearer. For confirmation on the species, contact specialized paleontologists, geologists, and local academics. And if you're lucky enough to find one, a biostratigrapher (there are not enough of those in the world). "A species is not just a group of morphologically similar individuals, but a group that can breed only among themselves, excluding all others," noted Ernst Mayr, one of the leading twentieth-century evolutionary biologists. "When populations within a species become isolated by geography, feeding strategy, mate choice, or other means, they may start to differ from other populations through genetic drift and natural selection, and over time may evolve into new species. The most significant and rapid genetic reorganization occurs in extremely small populations that have been isolated."

"We know today that it is not that easy," noted prominent twenty-first-century paleontologist Christian Klug.

**Tropites subbullatus (Hauer, 1849)**
Order: Ceratitida. Appearance and lifespan: Late Triassic, 232 to 222 MYA. Location: Hallstatt Limestone, Vorder Sandling, Austria. Size: 4 cm.

## Ask Your Fossil Friends

Wise words from associates and experts allow one to heave a sigh of relief when it comes to cross-correlating species of the same age from separate locations. In our adventure, the specimen that most resembled the *Tropites subbullatus* was preserved as a small white calcite whorl on a gunmetal-gray stone that was once a muddy seabed. But is the ammonite from the Hosselkus Formation the same species as the one that has been identified from Germany's Hallstatt Formation?

And so, we ponder the eternal question, what have we found? Yes, it was discovered in the *Tropites subbullatus* zone, and that species of ammonite is characterized by a distinctive inflated-looking shell with a central keel and a deep umbilicus, which may be ornamented with delicate ribs. But without a comparison of the suture patterns and a microscopic inspection, you can only deduce.

## Paleo Specialties

Paleontologists who are affiliated with museums and universities often have the opportunity to progress their careers with research grants. Having a funding cushion for study has allowed scientists to focus on fascinating specialty areas. Sometimes this leads to discoveries that alter history as we know it—like the twenty-first-century realization that dinosaurs did not go extinct; rather, they led to the evolution of birds. A researcher at a major Midwestern museum that specializes in primordial fires as a way to study extinction patterns. A post-doctoral student at an Ivy League institution was fascinated with what trilobites ate—even though these ancient arthropods went extinct more that 252 million years ago. A world-renowned acquisitor has proposed a thesis with the working title, *Sexual Dimorphism among Trilobites; Prove Me Wrong, Sucker.*

Whoever it is, it's good to have paleo pals and fossil friends. We touched base with paleontologist and digger Andreas Spatzenegger; we purchased a *Tropites subbullatus* from the Hallstatt Limestone from him, so his opinion was helpful and enlightening. He noted, "On your pics I do not see any clear *Tropites*. The whitish button could be the rest of a former 3D preserved ammonoid. Most probably *Proarcestes* sp."

Identifying ammonites can indeed be a challenging, time-consuming task. But on the upside, it's tremendously satisfying to know what you have found. Plus, it adds provenance and increased value to your collection.

**Ammonite Bite:** Until identification is confirmed, you can identify a specimen by the genus, followed by (*sp*), for example, *Tropites (sp)*.

**_Spirogmoceras shastense_ (Smith, 1904)**
Order: Ceratitida. Appearance and lifespan: Late Triassic, 237 to 222 MYA. Location: Hosselkus Formation, Lake Shasta, California, United States. Size: 2 cm.

(right) **_Aspenites laevis_ Welter, 1922**
Order: Ceratitida. Appearance and lifespan: Early Triassic, 251.2–247.2 MYA. Location: Noe Bihati, Timor, Indonesia. Size: 3 cm.

*Girtyoceras* **Wedekind, 1918 (sp)** split.
Order: Goniatitida. Appearance and lifespan: Middle Carboniferous, 345 to 304
MYA. Location: Errachidia Province, Drâa-Tafilalet Region, Morocco. Size: 6 cm.

# Death and Fossilization

Ammonoids were like early scuba divers, able to use their buoyancy control to navigate their position in the water column and swim freely through the seas. They piloted ecosystems that were typically unable to sustain more inactive animal life. When you find a cluster of ammonoids, consider that they were possibly the last creatures to survive in poorly oxygenated waters, slowly suffocating into the big sleep. Ironically, these low-oxygen environments were sometimes an ideal locale for ammonoids to become embalmed in fine sea silt, clay, or marl and quite naturally begin the lengthy process that would eventually lead to their fossilization.

After death, water seeped into the chambers of the phragmocone, and the animal sank to the seafloor. Not unlike rusting metal, bacterial decomposition altered the oxidation-reduction conditions around the animal, and the conch became embalmed by sands, muds, and minerals. Under certain conditions, the ammonoid developed its own tomb, known as a concretion. These nodules allowed for excellent shell preservation and fostered a highly collectible variety of fossilizations.

## What Is a Fossil?

All that is left of *Ammonoidea* and other extinct fauna is the evidence of their existence within the fossil record. From primordial mudslides to endemic ammonoid environments, paleontologists garner

***Eleganticeras elegantulum* (Young & Bird, 1828)** split.
Order: Ammonitida. Appearance and lifespan: Early Jurassic, 182.7 to 174.1 MYA. Location: near Black Ven, Charmouth, West Dorset, England, United Kingdom. Size: largest is 4 cm.

their information from any vestige of preserved remains, impressions, and remnants of once-living entities from a prior geologic age. These illuminating fragments of epochs past vary in many different ways: location, age of preservation, mineralization, creature, environment; the list goes on. Take size, for example. Scientists have managed to isolate Jurassic and Triassic nanometer-tiny bacteria in the sandstones of Queensland, Australia. Halfway across the world, in northeastern Arizona, are the rock-hard, life-sized trees of the 200-million-year-old Chinle Formation. And then there are the dinosaurs. Scotty, the mighty Cretaceous *Tyrannosaurus rex* skeleton at Canada's Royal Saskatchewan Museum measures 42 feet long and weighs 9.7 tons. It was quite the project to excavate it from the province's Frenchman River Valley.

"Fossils are the remains or traces of ancient life that have been preserved by natural processes, from spectacular skeletons to tiny seashells. Imprints, tracks, and trails can also become fossilized, like dinosaur footprints or worm burrows. These are called trace fossils," explained a spokesperson for southern England's Jurassic Coast Trust. "By studying the remains of life and the traces it left behind, we can learn a lot about how animals and plants lived and behaved millions of years ago."

How a living thing becomes a rock is amazing, even in concept. One minute it is alive and in the slip of a mudslide, a spew of a volcano, or some other cataclysmic event, the fauna becomes entombed, smothered, and stuck until it is discovered in a later stage, if at all. The small percentage of active fauna that reach fossil immortality tend to be preserved in sedimentary rocks like sandstone, limestone, chalk, clay, and shale. These formations are often an aggregation of small granules coagulating around minerals or organic particles. Aided by some type of water source, the residues cement together and develop into rock.

While sedimentary dispositions are common on Earth's surface, they are but a minor component

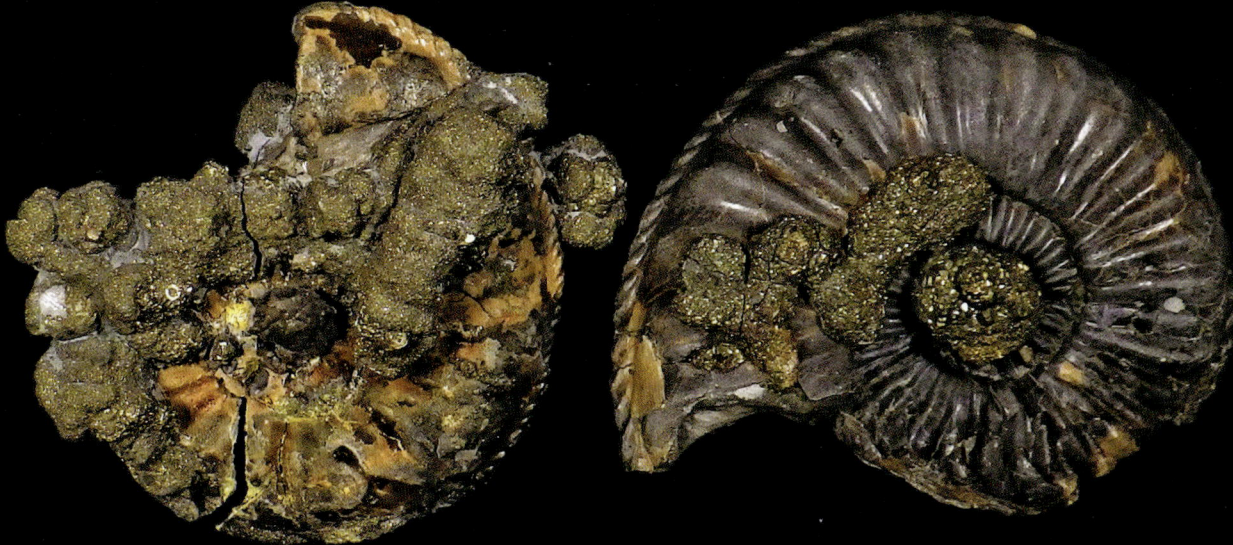

*Amaltheus* ammonites have a sharp keel that resembles twisted rope. These specimens are festooned with pyrite bursts.

**Amaltheus Montfort, 1808**
Order: Ammonitida. Appearance and lifespan: Early Jurassic, 192.9, to 184.2 MYA. Location: Zell unter Aichelberg, Germany.
Size: largest is 6 cm.

of our planet's crust, which is dominated by igneous and metamorphic rocks. (Sedimentary rocks have also been found on Mars.) The word "fossil" comes from the Latin word *fossilis*, meaning "dug up." Thus, a creature that became entombed in stone officially becomes a fossil when it is excavated.

**Ammonite Bite:**
- Reduction is the gain of electrons.
- Oxidation is the loss of electrons.

## How Old Is a Fossil?

Preservations are usually considered to be fossils if they are more than ten thousand years old. That allows still-active locations like the La Brea Tar Pits in Los Angeles to be included in this elite group. On the other end of the spectrum, some microfossils discovered in Australia are thought to be between 3.48 and 4.1 billion years old.

Science is open-minded about what can be considered a fossil. If the find is old and provides information about life past, the academic community usually approve of the designation. The term covers a wide spectrum, as you've likely noticed, and the process of fossilization varies tremendously. Two of the main criteria are tissue type and external conditions. Typically, only a portion of the deceased organism is preserved, and that segment was often already partially mineralized during the creature's lifetime. Calcium is a key for vertebrate preservations, which is why bones and teeth are among the most common animal parts found. With molluscs and other invertebrates, aragonite and calcite are the magic crystals for shell creation, as well as other protective exoskeletons.

Rusting iron is an example of a redox reaction. When an iron metal is exposed to atmospheric oxygen in the presence of moisture, it forms iron oxide (that is, rust), a common component of many fossils.

Image credit: Anna Frodesiak, CC0, via Wikimedia Commons.

The preservations from the La Brea Tarpits are among the youngest conservations to be labeled fossils. Between fifty thousand to eleven thousand, years ago, the sticky bog was covered with a rainwater sheen, the mirage of a quicksand oasis. Vertebrates got stuck and perished in the mire. To date, more than 1 million, bones have been excavated from the midtown Los Angeles site, including some four hundred dire wolf skulls.

*Aenocyon dirus* **(Leidy, 1858)**
Order: Carnivora. Appearance and lifespan: Late Pleistocene, 125,000 to 9,500 MYA. Location: Los Angeles, California, United States. Size: Average is approximately 30 cm.

## Redox Reaction

On its sea bottom deathbed, an ammonoid instinctually assumed its mortality position, drawing the soft body into its conch with its arm tips creating a mortal crown around its aperture. As the non-biomineralized parts of the animal decayed, the creature's body released a science experiment of elements and gasses. (It probably stank up the area; aromas can be smelled underwater.)

This finite reduction-oxidation reaction could be augmented by bacteria or stimulated by scavengers, aiding or eliminating an ammonoid's body decomposition. Over time, the fossilizing shell absorbed phosphates, carbonates, and/or other minerals suspended in the seawater. Known as a redox reaction, this common chemical consequence changes the oxidation states of atoms by transferring electrons from one atom, ion, or molecule to another particle with the same chemical formula.

Many reactions in organic chemistry involve changes in oxidation states. These transitions may occur slowly, as in the formation of rust or the fossilization of an ammonoid. An example of a rapid redux reaction is burning fuel. Oxidoreduction is critical to the circle of life. Our bodies use redox reactions to convert our intake of food and oxygen into energy plus water. When we exhale, we contribute to the carbon dioxide already present in

The Republic of Madagascar is recognized for some of the most attractive ammonite preservations on the planet. This **Cleoniceras** has chambers conserved with lemon calcite. The island is renowned for its extraordinary array of eye-catching minerals such as celestite and rose quartz.

*Cleoniceras* **Parona & Bonarelli, 1897** (sp) split.
Order: Ammonitida Appearance and lifespan: Early Cretaceous, 113 to 100.5 MYA. Location: Ambarimaninga, Mitsinjo, Boeny, Madagascar. Size: 11 cm.

For ammonoid preservation, aragonite typically dominates the fossilization process, but it depends on marine chemistry and other environmental factors. "Aragonite is best suited in warmer and shallower water. It does not dissolve easily in such conditions, thus, it is the preferred composition for most 'seashells.' In colder conditions, and deeper water, calcite is needed for the shell to survive," explained Neal L. Larson, a self-described geologist, paleontologist, collector, preparator, researcher, author, and confessed ammonite addict.

On land, aragonite eventually transforms into calcite unless it is somehow stabilized. Modern mollusc shells like nautili and snails as well as bivalves are made almost exclusively of aragonite.

**Ammonite Bite:** The process by which a living creature turns into stone is known as diagenesis.

the atmosphere and thereby help the flora to create redux reactions of their own.

In the water world, when things are in the process of preservation, the redox reaction of a decaying corpse becomes the catalyst for a host of biochemical reactions. The *Ammonoidea* fermentation process aggregated an assortment of sea resources that may have included concentrations of volcanic ash or tectonic plate–induced mud. Sometimes this facilitated the development of a globular concretion forming around the animal. It is this muck and mineral aggregation has been attributed to the outstanding preservation of many ammonoid fossils.

During decay, minerals may replace all parts of the animal's shell.

*Quenstedtoceras* **Hyatt, 1877** pyrite split.
Order: Ammonitida. Appearance and lifespan: Middle Jurassic, 164.7 to 161.2 MYA. Location: Saratov Oblast, Russia. Size: 10 cm.

## Permineralization

Of all fossils, cephalopod shell preservations offer some spectacular collection-worthy specimens. The most typical method of ammonoid fossilization is known as permineralization. Conch conservation was a relatively simple scientific process whereby the aragonite is gradually recrystallized into calcite. This element-rich process has been known to create cephalopod conch preservations revealing extraordinary chamber and suture patterns. "For soft tissue preservation, bacterial activity sometimes helps to precipitate calcium phosphate in and around soft parts," elaborated Christian Klug. "This is a very complex process, which is not fully understood in all details."

Of course, the result of any diagenesis depends on the loose sediments and minerals suspended in the surrounding environment, the chemical alteration conditions, and the compaction method. When you ponder all the circumstances necessary for fossilization, it is incredible that Earth has left behind so many exceptional evolutionary clues.

**Ammonite Bite:** Unless stabilized, pyritized ammonites may decay over time. Heat and moisture contribute to this action. Known as pyrite disease, it is another form of oxidation.

Beyond shelled sea creatures and other marine organisms, permineralization has preserved plants, teeth, and bones. Famous examples of this type of preservation are the Late Triassic tree trunks from the Petrified Forest in northeastern Arizona. The organic materials in these logs have been replaced by silicates such as chalcedony and quartz, leaving the cellular structure of each tree intact. Their rich brick-red color comes from iron oxide; the ochres, purples, grays, and blacks were aggregated from other minerals.

Pyrite is another compound that regularly makes its way into ammonoid preservations. In this scenario, as the organic matter decayed, it released sulfides that interacted with the iron ions

*Harpoceras falciferum* **(Sowerby, 1820)** with *Dactylioceras* **Hyatt, 1867** Order: Ammonitida. Appearance and lifespan: Early Jurassic, 201.3 to 145 MYA. Location: Holzmaden, Baden-Württemberg Region, Germany. Size: 25 cm.

dissolved in the surrounding waters. The conditions ultimately allowed pyrite to invade the shell and fill in the chambers, creating a preservation with a brassy luster. It is possible to find exceptional metallic-looking specimens at various locations around the world. Pyrite-replaced *Quenstedtoceras* ammonites from Ulyanovsk, Russia, are often sliced and polished to enrich their decorative value. Ammonoids and goniatites of Europe and North Africa have also been known to be replaced by silver pyrite. If preservation takes place in a high-temperature, oxygen-rich environment, the replacement material may become hematite. Should conditions cause the pyrite to rust during the fossilization process, the replacement material might become limonite. These iron oxides tend to be more stable than golden pyrite and marcasite specimens.

## We Dig Display Ammonites: Holzmaden, Germany

Deutschland is an ammonoid seeker's paradise. Throughout Germany, extraordinary locations like Solnhofen, Bundenbach, Hunsrück, and Messel, among others, offer exceptional prospecting opportunities. So fine are the country's collecting prospects, the label Lagerstätte ("storage site") was introduced into the paleontological lexicon in 1970 by internationally recognized Professor Adolf Seilacher. It is defined as "fossil deposits in bodies of rock that contain an unusual amount of paleontological information in terms of quality and quantity." Conservation folklore has it that the term was coined because of the magnificent preservations at Holzmaden. Some of the most picturesque fossils are the pyrite-highlighted tableaus from this Lagerstätte. The dark oil-shale matrix captures an Early Jurassic portrait of the sea shelf that once covered Central Europe. German paleontologists christened Holzmaden's distinguished formation

the Posidonienschiefer, named after the oyster-related bivalve found throughout this extensive conservation. (It has recently been renamed the Sachrang Formation.)

Meandering through sections of Germany, Switzerland, Austria, Luxembourg, and the Netherlands, the Sachrang Formation is renowned for its telling preservations of marine invertebrates, fish, and reptiles. The deposit began some 180 million years ago, when the climate consisted of summer monsoons and dry winter trade winds. The combination caused notable seasonal fluctuations in the oxygen content of seawater, resulting in a petrified sequence of finely laminated layers of oil shales formed of fine-grained sediments intercalated with bituminous limestones.

Dead creatures were captured in an anaerobic environment, and they only gradually decomposed, leaving an exceptional variety of the Early Jurassic's top predators—reptiles like crocodilians, plesiosaurs, and ichthyosaurs. They plundered the area fish, including—rare sharks and coelacanths. The area's conserved invertebrates are numerous and astounding, with wondrous squids and crinoids, and more than a hundred species of ammonites.

The Sachrang Formation crops out in a number of locations in southwestern Germany. But because Holzmaden's fine preservations may look like art, it is a go-to spot in the formation's shale. This unique stratum was deposited on the seafloor during the early Toarcian Stage, when the Tethys Sea was connected to the open ocean through continually narrower routes. Consequently, freshwater supplies grew limited, creating an anoxic deepwater environment. This low oxygen environment was compounded by algae blooms near the surface, which in turn caused some areas to build up toxic hydrogen sulfide and bitumen on the seafloor. Because there were no aerobic bacteria to eat away the skin and tissue, and the water was too inhospitable for scavengers, corpses and conchs were covered by settling particles. The deposition sludge was rapidly compressed by the raining sediments. These circumstances allowed the flattened remains of the animal to become stamped in stone. So detailed are Holzmaden preservations that organs, skin, and muscle may have carbonized, or the skeleton became permineralized—infused with pyrite and phosphate.

During the Early Jurassic, the Tethys Sea was a paradise of Ammonoidea, and many fine examples may be found at Holzmaden. Whether you are digging or acquiring ammonites, look for any of a score of members from the suborder Lytoceratina. This grouping is identified by their tapered, loosely coiled, evolute shells and complex sea-fan suture patterns. Also in the mire are several dozen members of the superfamily Hildoceratoidea. *Hildoceras* was a characteristic species of this largely evolute group. This genus typically had a narrow outer axis (also known as a venter) that may or may not be divided by a sharp keel. Their test ornamentation ranged from a symmetrical line design like a wave to deep, thick ribbing. Nearby may be family member *Mercaticeras*, an evolute genus with picket fence struts and squarish tubing that was sometimes wider than tall. The score or more of Dactylioceratidae family members had a shell reminiscent of a compressed tube with springs wound around each coil. The venerable Phylloceratidae was also present at Holzmaden. The name is from the Greek words "leaf horn," as their sutures have been said to resemble oak leaves. The half dozen or so species found in this shale featured thin-walled, smooth, involute shells.

Men of science have been making records about Holzmaden since Duke Carl Eugen of Württemberg documented the area in 1749. Soft body tissue was discovered nearly 150 years later when the renowned preparator Bernhard Hauff noted fossil remnants of soft tissue while preparing an ichthyosaur in 1892. His family founded the Urweltmuseum Hauff (Hauff Museum of the Prehistoric World) to honor his collection and his pioneering excavations, preparations, and observations.

Regional residents and merchants have been accessing the slate quarries around Holzmaden since Medieval times. Modern-day quarrying of the meters=thick Posidonia Shale focuses on a narrow tier of rock called the Fleins Layer. The shell-encrusted stone is used for interior design—countertops, flooring pedestals, and the like, and for exterior protection such as on roofs and walls. When the quarry workers come across fossils, they are then excavated and salvaged by museum specialists. Large, floating sea lily crinoids and the vast majority of area vertebrates (except crocodiles) come from this layer.

For private collectors and commercial digging, try the Kromer slate quarry in Ohmden, located 2.5 kilometers (1.5 miles) from the Urweltmuseum Hauff museum. It is a well-managed site where new layers of shale are regularly exposed, offering optimal conditions for finding many species of ammonites as well as bivalves, belemnites, pyrite, fossilized wood, fish scales, and the remains of crinoids.

Both the Urweltmuseum Hauff and the fossil pit, Schieferbruch Ralf Kromer, are accessible from the autobahn A8. Take the Aichelberg exit, which is about two hours northwest of Munich and a half hour southeast of Stuttgart. The quarry charges a minimal entry fee. Hammers and chisels are available for lease and definitely add to the experience. It takes a little time to train your eyes to spot the fossils, which may look like little more than bulges in the rock. But once you get the hang of it, you'll likely come away with as many fossils as you can carry. Professionals prepare Holzmaden fossils using sandblasting tools to remove the matrix from vertebrae, and detailed work is done with the use of a binocular microscope.

## Replacement Materials

Calcite and aragonite may also be replaced by a textural variety of silicates, including chalcedony, opal, agate, or even quartz. *Turritella Agate* is the name given to the gemmy, brownish matrix material encasing Eocene snails at the Green River Formation in Wyoming. The torpid region of Coober Pedy, Australia, supplies much of the world's gem-quality opal, as well as the occasional fossil, such as belemnites and clams. At this Early Cretaceous location in the southern part of the continent down under, debris from ancient volcanos

*Dactylioceras* ammonite group.

*Dactylioceras* **Hyatt, 1867**
Order: Ammonitida. Appearance and lifespan: Early Jurassic, 183 to 182 MYA. Location: Holzmaden, Baden-Württemberg Region, Germany. Size: largest is 12 cm.

seeped into and replicated the remains of a formerly living creature that was buried in the mire. The conditions have turned the silicate into brilliantly colored casts known as belemnite pipes and pearlescent bivalves.

The ancient, petrified trees of Puyango relate a story of sea level rise. They stand (or lie, as the case may be) as some of the largest single fossil organisms ever produced by our world.

Image credit: Arabsalam, CC BY-SA 4.0, via Wikimedia Commons.

## We Visit Ammonites: Mixed Mediums in Ecuador

**Name: Bosque Petrificado de Puyango**

**Location: Ecuador**

**Age: Cretaceous**

**Date of UNESCO Geopark Inscription: January 2015**

### How Puyango Got to Be

The impressive peaks of the Andes form a continuous highland along the western edge of South America, creating the longest continental mountain range in the world. About 96 to 112 million years ago, they were built by the continuous tectonic activity that macerated everything in its wake. Fire and molten rock spewed from the cracks, while lava bombs flew in all directions, charring wherever they fell to Earth. In turn, superheated ocean waters flooded the terrain, and the woodland slowly became submerged. Silicification and carbonization preserved hundreds of trees rich in paleontological information. This cordillera has experienced regular geomorphic activity since the breakup of the first supercontinent, which is why we are gifted with the fascinating petrified forest at Puyango, a paleontological area spotted with ammonite locations.

**Ammonite Bite:** A UNESCO geopark is a physically special topographic area where the landscapes and notable sites are protected from development and shared for their splendor and to educate visitors about the surroundings. Geoparks expect the participation of local communities to achieve United Nations conservation objectives.

**Ammonite Bite:** The Puyango Petrified Forest (PPF) is one of the few sites in the world where you can analyze the paleontological aspects of the prehistoric flora and relate it to the plant landscape existing today.

*Buergliceras buerglii* **Etayo Serna, 1968**
Order: Ammonitida. Appearance and lifespan: Early Cretaceous, 129.4 to 125 MYA. Location: Paja Formation, Villa de Leyva, northeast of Bogotá, Colombia. Size: 7 cm.

Dwarfed by the massive tree trunks, ammonites may be spotted in the paleontologic melee upstream at Las Concreciones. In the sedimentary outcrops, one can locate internal and external molds of these cephalopods as well as smaller bivalves. Petrified specimens may also be identified by other streams in the formation, including the Cochurco, El Chirimoyo, El Limón, and El Guineo.

## Geology

This unique paleogeographic landscape developed during the Cretaceous Period, when tectonic subduction went on beneath parts of Peru but left the Ecuadorian area submerged by the Pacific Ocean. As the region evolved, the territories that were bathed by seawater eventually dried out, allowing the region to become populated with forests and animals. As millennia passed, mountains rose, and some of the terrestrial life was buried during the process. The persistent geomorphic activity also pushed marine critters like ammonites, echinoderms, and bivalves to the surface, creating an invaluable sample of both the land-dwelling and oceanic past of the planet.

## Trees

The trees are the main attraction of the Puyango Petrified Forest. Preserved in stone, they're an extraordinary site—particularly the *Araucarioxylon* petrified cone pines. This geopark began its time on the planet as a gymnosperm timberland growing on a relatively level area. Today it rests in a rather narrow basin shaped by geomorphic activity.

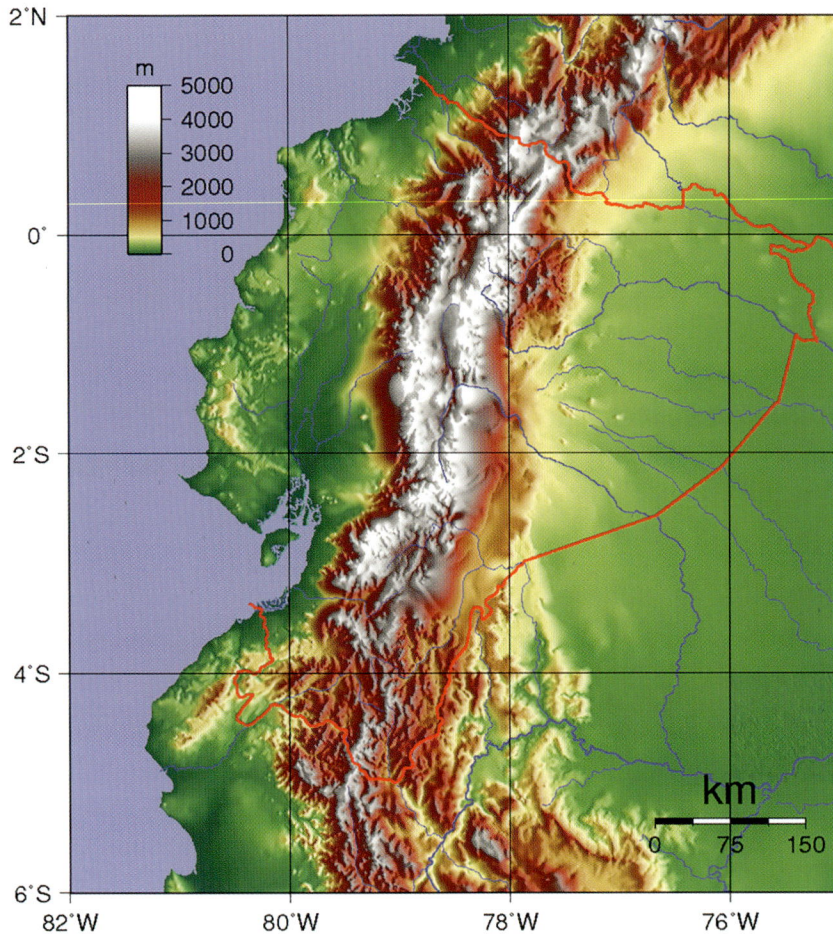

This topographic map of Ecuador also conveys the longitude and latitude of this South American equatorial country.

Image credit: Sadalmelik, public domain, via Wikimedia Commons.

Located in the south of Ecuador, the Puyango stone forest is now elevated well above sea level at a height that tops most high-rise skyscrapers. It covers an impressive 2,659 hectares (6,571 acres) and straddles the provinces of El Oro and Loja, Puyango, and expanding beyond the border region into Peru. Ultimately, more than 500,000 hectares (1,235,527 acres) are protected by this geopark.

"The Puyango Forest is one of the few petrified forests in the world, which makes it of great importance because the history of humanity and life in that territory is recorded," observed local politician Hernán Encalada. These Cretaceous conifer conservations are impressively large. Stone tree trunks may have a diameter as wide as 2 meters (6.6 feet), and stretch as long as 15 meters (49 feet) as they lay across the forest floor. There are also four types of preserved leaves, indicating ancient versions of ferns and palms. The quantity of petrified trees contained in such a compact area represents one of the largest collections of this type in the world, comparable to Arizona's Petrified Forest National Park in the United States.

**Ammonite Bite:** This area is an elaborate maze of low mountain ranges dispersed with valleys and plains, offering a significant contrast to the rest of the Ecuadorian Andes. In this southern region, the cordillera drops down to mere hillocks compared to the peaks that tower 2,800 meters (9,186 feet) above sea level.

## Ammonites among the Trees

Not too young, not too old, the ammonites uncovered at the Puyango Formation were born in the heart of the Cretaceous Period, some 108 million years ago. These cephalopods are found intermingled with petrified wood in the upper part of the area, which is a veritable coquina of fossils that have coalesced in an ash matrix. This curious intermingling of terrestrial life and sea creatures came about when tidal and geothermal activity forced tree trunks into an aqueous medium of seawater and volcanic residue. Fossils became rocks, and wood turned to stone, and this curious combination resides in situ. It is indeed a special occurrence to find ammonites hidden among fossilized trees.

The cephalopods found in this area are not the real thing but an incredible simulation. Rather than being the ammonites themselves, the Puyango specimens are casts of shells. In the space where the creature came to rest, the conch was replaced by both internal and external molds of saturated aggregates of minerals. These coveted specimens are found in the vicinity of several streams, including Cochurco, El Chirimoyo, El Limón, and the appropriately named Las Concreciones.

As you are poking around for rocks in the mud, look for nodules with the thick-ribbed *Dufrenoyia.* There is also the remarkably interesting *Peltoceras,* with its exposed coils that lightly touch; a wide, flat outer rim; and notable ridges that bifurcate close to a decorative edge. In addition, expect to find the involute *Schloenbachia,* which is thickest near the umbilicus; has gaps between ridges; and a wavy, noded outer coil. It grew to the diameter of a billiard ball and had a delicately leafed suture pattern. Much larger was *Forbesiceras,* with an involute conch that may have a diameter larger than a soccer ball. A descendant of *Schloenbachia,* it is notable for its smooth narrow shell that covers the

These silicified natural internal molds capture the details of the original ammonite shells.

From top right: Likely ***Mammites nodosoides* (Schlüter, 1871)**, ***Choffaticeras (Choffaticeras) segne, Mammites nodosoides, Wrightoceras munieri* (Pervinquiére, 1907)**
Order: Ammonitida. Appearance and lifespan: Late Cretaceous, 93.5 to 89.3 MYA. Location: Errachidia Province, Tafilalt Region, Morocco. Size: largest is 4 cm.

**Ammonite Bite:** Coquina is a sedimentary rock largely composed of sea and shell fragments of ammonoids, gastropods, ostracods, and bivalves lightly cemented together. It may contain silt, ash, and clay particles, thus making it an extremely porous rock. Coquina sometimes serves as an aquifer or a reservoir holding paleontological secrets. If calcium carbonate precipitates within the deposit, it likely forms a cement that binds the fossil debris together, transforming into a sedimentary rock.

**Dufrenoyia furcata (Sowerby, 1836)**
Order: Ammonitida. Appearance and lifespan: Early Cretaceous, 122.46 to 112.03 MYA. Location: Kreide Unit, Guane, Colombia. Size: 5.5 cm.

A serene sunset over Peru's Napo River.

Image credit: Murray Foubister, CC BY-SA 2.0, via Wikimedia Commons.

more mature whorls. It begins with a closed tight center and terminates with sharp rims.

When to visit Ecuador's grove of petrified woods depends on what you want to experience. The subtropical climate boasts a comfortable average temperature of 22.5°C (72.5°F). Exploring is easiest during the dry season, which generally runs from May to December. During the winter season, the area receives more than 900 mm (35 inches) of rain, which, as you might imagine, considerably changes the landscape. During the rainy periods, waterfalls pour off the cliffs, while streams course through the landscape, invigorating the vegetation of its ecosystem. The downpour brings new life to Ecuador's dry forests, and locals delight in the magnificent golden blooms of the Guayacanes de Zapotillo woods, which flower in the early part of each year.

## We Dig Ammonites: The Republic of Ecuador

The Andes fault line has created many ammonite-rich locales around this area, originally renowned as the Inca Empire and later conquered by Spain. Look for Cretaceous cephalopods along the muddy Rio Napo, a tributary to the Amazon River that crisscrosses the Oriente—everything east of the Ecuadorian Andes. By land area, the Oriente is about half the size of the country, much of it virtually unexploited tropical forest. The area is inhabited by a tiny fraction of the country's population, who live mostly in small villages along the river. It's an excellent location for an ammonite-driven exotic adventure.

Make your way to Puerto Misahualli and ask around for the sections where marine fossils are abundant. You will find ammonites huddled along with bivalves, echinoids, and microfossils. There are many beds where preservations lie like Sleeping Beauty awaiting the humble kiss of an explorer's rock hammer. The locals will let you know where to dig for specimens; people at museums and colleges are good for mining details. They'll let clue you in on which pockets hold the notably ridged and finely sutured *Brancoceras*, recognized by its rounded venter. Perhaps they'll share where to find

*Peroniceras*, which was ribbed with many whorls vaguely resembling an ancient wheel. *Barroisiceras* was large, corrugated, noded and has a keel that looks like the fin on Aquaman's head. There are other fabulous ammonite finds like *Mortoniceras*, *Neophlycticeras*, and *Dipoloceras*. What you discover depends on who you ask and where you look.

## Fossil Preservation

The preservation process that turns an animal or plant into a fossil is known as taphonomy. The remarkable procedure that takes what was once alive and turns it to stone involves three main processes: necrosis, biostratinomy, and diagenesis. These are fancy words for an easy to comprehend sequence:

1. Necrosis is the death of living tissue.
2. Biostratinomy covers the course of events from death to burial.
3. Diagenesis is the transformation of soft sediment or other deposits to rock.

"Necrosis" is taken from the Greek word for "death." It sounds like other terms related to mortality—such as necromancy and necrophilia. Thus, it is easy to associate the term "necrosis" with the death of living tissue.

**Mortoniceras Meek, 1876**
Order: Ammonitida. Appearance and lifespan: Early Cretaceous, 112.03 to 109 MYA. Location: Fort Worth Formation, Tarrant County, Texas, United States. Size: 9 cm.

"Biostratinomy" is a blanket term for the processes that take place after an organism dies but before its final burial. The time for this progression can vary from the second the tidal wave hits to the minutes it takes to get engulfed in amber, to the many years necessary to accumulate a mass casualty bone bed at the base of a cliff.

"Diagenesis" refers to the physical and chemical changes in deposits that occur under rising

A Jurassic fossil cluster with ammonites, coral, brachiopods, and crinoids.

Order: Ammonitida. Appearance and lifespan: Middle Jurassic, 165.3 to 161.5 MYA. Location: Fernette, South Alsace, France. Size: largest is 2 cm.

temperature and pressure; the process that yields a fossil. It usually involves lithification—the preservation building block in which sediments compress, banish fluids, and eventually solidify into rock. This progression allows minerals to become a replacement material for shell, wood, bone, and non-biomineralized matter.

For a living thing to turn to stone sounds like an ancient myth, but we know it is a reality. Of course, plenty of events can go wrong during diagenesis, allowing the fauna or flora to erode away and be forever forgotten. Too much heat can turn organic molecules into oil and gas. Crushing a critter in coarse sand can fragment the cuticle that protects the animal from the environment, and likely end all possibilities of fossilization.

## Deposits

There are essentially two underwater processes that lead to exceptional preservation clusters: stagnation and obrution. Stagnation deposits occur primarily in open sea communities that succumb to low oxygen conditions in torpid or hypersalinated bottom waters. Stagnant anoxic seas allow for the metaphoric pickling of fauna. This dormant stasis delays the decomposition of biological features until long after a durable impression has been etched in the surrounding matrix. The Posidonia Shale of the Holzmaden formation in southern Germany is a fine example of Jurassic preservation by stagnation. The coal-gray marls preserve marine life at distinct intervals, earmarking the oxygenation cycles of the seafloor.

*Dactylioceras semicelatum* **Simpson 1843** plate with hand painted identification.

*Dactylioceras semicelatum* **Simpson 1843**
Order: Ammonitida. Appearance and lifespan: Early Jurassic, 183 to 182 MYA. Location: Posidonia Shale, Holzmaden, Baden-Württemberg Region, Germany. Size: largest is 6 cm.

On a larger scale are obrution deposits, where sporadic smothering by sand, mud, or clay sea sediments expedites burial. The well-known Middle Cambrian Burgess Shale conservation in the Canadian Rockies is a quintessential example of preservation by obrution. Evidence indicates that bottom-dwelling communities were periodically pressed to death in landslips. This conserved a rare and treasured glimpse of the weird wonders that lived in an ecosystem that is approximately 508 million years old. The more recently uncovered Chengjiang biota was alive even earlier: it is about 518 million years old. The early to middle Devonian trilobite beds of Morocco were entombed in episodic obrution beds in a way that it now records Paleozoic changes in ocean climate.

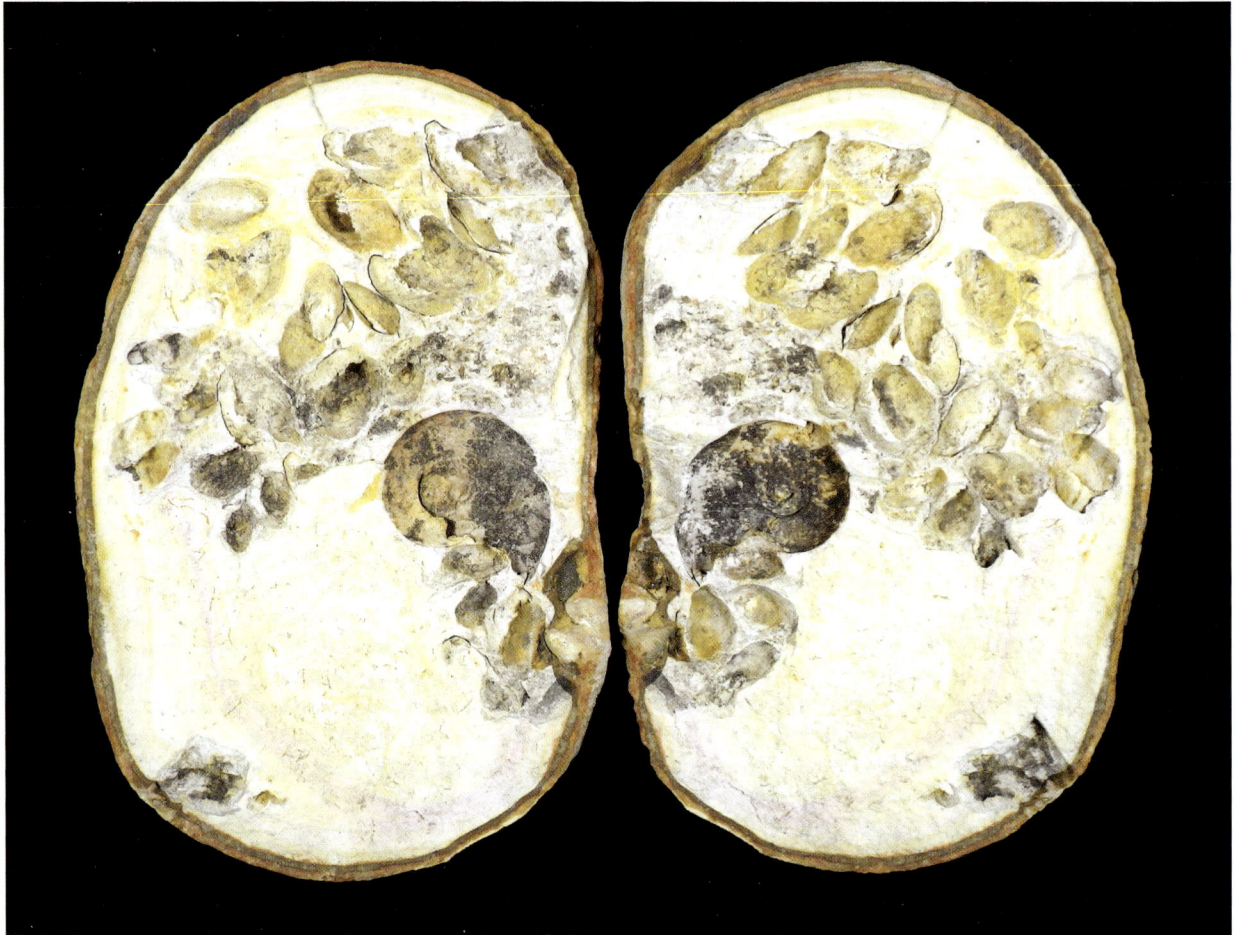

This Triassic ammonoid and bivalve concretion captures a once thriving community that may have been buried by pulses of mud-rich sediment.

## The Process

As you would likely expect, so many variables affect the processes that lead to fossilization, which results in a wide variety of preservations, even among the same species. Some of the diverse types include:

- Adpression
- Casts and molds
- Permineralization
- Mineralization
- Replacement
- Recrystallization

When you crack open a concretion and see the fossil on one side and the imprint of the ammonoid shell on the other, that is a specimen created by both compression and impression all neatly wrapped inside a concretion. If the fossil is a combination of these conservations, it's known as adpression—a distinctive type of cast-and-mold formation that may happen with concretions.

Cast fossils generally occur with creatures with shells that easily disintegrate or dissolve. During the process of authigenic mineralization, the carcass and conch slowly erode, forming an organism-shaped external mold in the rock. In turn, the

*Colchidites breistrofferi* **Kakabadze and Thieuloy 1991**
Order: Ammonitida. Appearance and lifespan: Early Cretaceous, 130.0 to 125.45 MYA. Location: Paja Formation, Santander, Colombia.
Size: largest is 5 cm.

cavity is filled with a mixture that includes the conch's decomposing aragonite. Voilà, the internal mold has become an endocast of the original.

Permineralization occurs when an organism is buried soon after death but prior to decay. Minerals precipitate from the sea, occupying spaces that were filled with liquid or gas during the ammonoid's life.

Aggregate minerals can seep into spaces even more delicate than septa walls. With ammonoids, this produces magnificently detailed fossils that may show suture patterns. Permineralization has also preserved plant cell walls, traces of skin, feathers, and even the soft tissue and organs of animals.

This *Cleoniceras* ammonite split reveals an undulating pyrite suture design.

**Cleoniceras Parona & Bonarelli, 1897** (sp)
Order: Ammonitida. Appearance and lifespan: Early Cretaceous, 113 to 100.5 MYA. Location: "Ammonite" bed, Ambatolafia, Boeny, Madagascar. Size: 5.5 cm.

# 6

# Calcite and Aragonite

Life in Earth's seas began approximately 4 billion years ago. Some 3.6 billion years later, several rogue freshwater plants edged their way onto muddy shores and swampy estuaries, marking the beginning of terrestrial life. There are eons of time between the start of life in the oceans and the existence of fauna on land, so the world's ancient marine environments are critical habitats for understanding the geological record.

Calcium carbonate ($CaCO_3$) mineralization is an important part of oceanic life and marine geochemistry. This common substance is the main component of pearls, bird eggs, and oyster and other shells, among other things we seldom take time to appreciate. In the form of stone, this compound regularly forms the minerals calcite and aragonite. Curiously, although aragonite and calcite have the same chemical formula—$CaCO_3$—their atoms are stacked in different configurations. This polymorphic molecular structure allows solid material to exist in more than one form. Aragonite builds slender needlelike crystals, whereas calcite forms blocky orthorhombic shapes. Which of the two aggregates facilitates ammonoid preservation depends on water chemistry and other variables.

Of the two, calcite is the more stable compound, although as temperatures and compactions change, one mineral may recrystallize into the other. In terrestrial settings, aragonite eventually transmutes into calcite. Undersea, where water pressure can be intense, aragonite is the preferred crystalline structure because it is the denser and typically

This impression of a *Placenticeras* ammonite picked up the ammolite shell when the nodule was opened. Dormancy, decomposition, pressure and an anaerobic environment combined to imprint the nacre into the surrounding matrix.

*Placenticeras* **Meek, 1876**
Order: Ammonitida. Appearance and lifespan: Late Cretaceous, 83.5 to 70.6 MYA. Location: Bearpaw Shale, St. Mary River, Alberta, Canada. Size: 7 × 8 cm.

**Ammonite Bite:** Polymorphs are two or more crystalline materials that contain the same chemical composition but differ in their molecular arrangement and crystal structure. Diamond and graphite are polymorphs of one another.

stronger. The transformation from aragonite to calcite (or vice versa) is referred to as recrystallization. More broadly speaking, this term is applied when the shell, bone, or other tissue is substituted by alternative elements. As with many ammonoids, mineralization of the original shell occurs so gradually that details such as suture patterns and other microstructural features are preserved, even with the total loss of the original material.

In the grand scheme of things, organisms are only rarely preserved as fossils. Figure that the number of species known through the fossil record is fewer than 5 percent of the actual amount of acknowledged living creatures. The varieties known through preservation thus may easily be less than 1 percent of all the fauna that has ever existed. And the fossil record is strongly biased toward organisms with bones, shells, and other hard parts, with most groups of soft-bodied organisms offering few clues for paleontology sleuths.

## Ammolite

Ammonoid collectors are known to be giddy about specimens preserved with ammolite. Those that acquire fossils for their aesthetic value delight in the iridescent hues that seem to glow from this unique precious preservation of ammonite shell. This special conservation can resemble an opalescent rainbow that shimmers and reflects color. The finest grade of gem quality ammolite is recovered along rapid watercourses in southern Alberta, Canada. Several long-established commercial mining operations are based along the banks of the St. Mary River. Between the town of Magrath and the city of Lethbridge is a vein of the Bearpaw Shale with ammonites that have been fossilized under optimal conditions for ammolite preservation. This is the only area known to yield this biogenic gem in commercial quality and quantity. "In this location, the amount of heat and pressure to preserve the aragonite was just right," explained John Issa, the sales and marketing director for Enchanted Designs LTD, a company that specializes in ammolite ammonites. The extraordinary conditions of the Cretaceous shoreline breeding ground at this locality preserved ammonites with remineralized and recrystallized aragonite intermingled with traces of iron, yielding ammonite preservations of the most extraordinary color and luster.

Half of all ammolite deposits are contained within the reservation of the Kainah First Nation Indian tribe (also known as the Blackfoot), allowing them to play a leading role in mining and the global marketing of the gemstone. "This stone is magic," noted commercial prospector Troy Knowlton, a First Nation tribe member. "In ancient times, my people held ceremonies where these rocks are blessed and painted. Women dance with the stones while the men sing the songs."

Inside the lode at the Bearpaw Formation, three ammonite species are found with ammolite preservation. The large involute *Placenticeras meeki* had a smooth shell. The involute *Placenticeras intercalare* had a ring of knobs on its flank. These disc-shaped species were accompanied by the tubular *Baculites compressus*, with its long, tapered shell. When these creatures died about 71 million years ago, they sank to the bottom and were buried by layers of bentonitic mud that eventually compressed into shale. This sedimentation process preserved the aragonite of the conchs, preventing it from converting to calcite. Instead, the ammolite was transformed into an opalescent biogenic gemstone.

**Ammonite Bite:** Ammolite is one of just a few biogenic gemstones; others include amber and pearl.

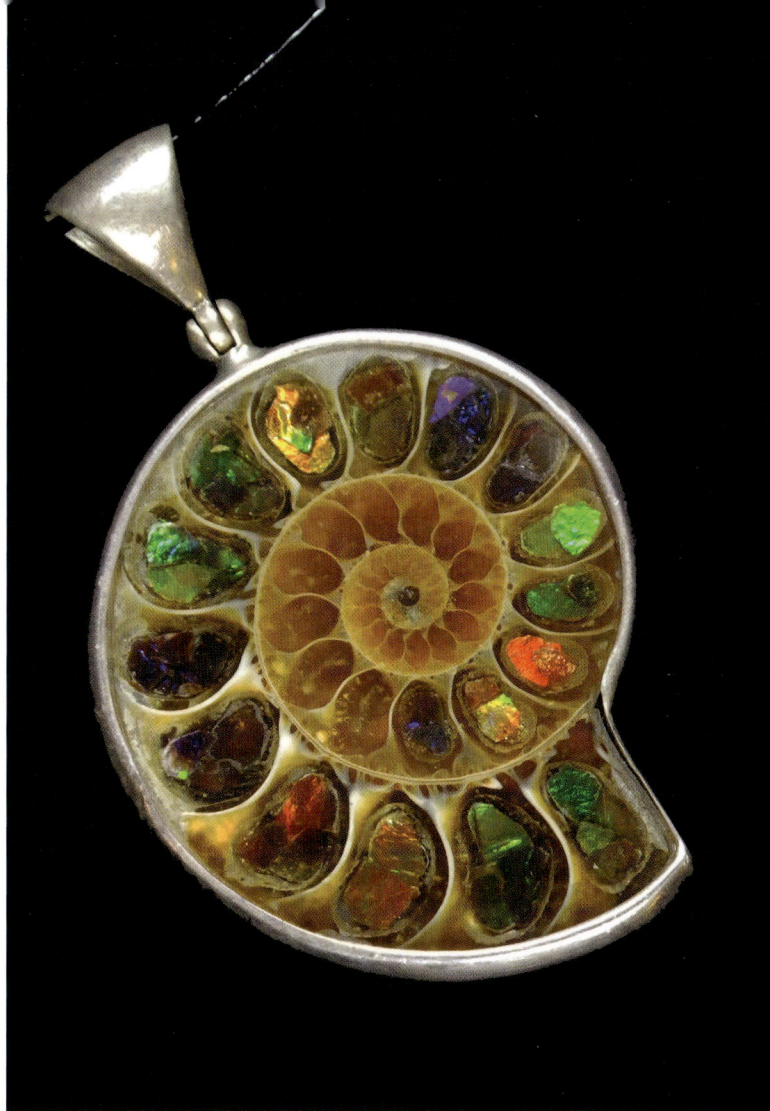

This cosmopolitan pendant features a **Cleoniceras** ammonite from Madagascar inlayed with Canadian ammolite.

Most gems get their color from light absorption and reflection Ammolite, on the contrary, gets its iridescent hues from light that refracts off the stacked layers of thin platelets that make up the aragonite. When freshly quarried, these ammonites' gemmy qualities are not especially vibrant, but they are fragile. "Out in the field, you have to go through a football field to find a shoebox worth of material," explained Issa. "Rock will cleave where the ammonite is if it's close to the edge. Once we spot this clue, the backhoe and heavy machinery stops, and everyone is on their hands and knees

A lone bison grazing along the plains of the Black Hills in South Dakota, United States.

Image credit: Idawriter, CC BY-SA 3.0, via Wikimedia Commons.

looking for the ammonite. When its located, you need to pour enough PaleoBOND™ fossil glue on it, so your eyes start to water and then walk away. Let it set for half an hour and then come back and take it out of the ground; otherwise it crumbles."

At the lab, ammolite requires polishing and preparation to expose the full potential of the colors. The thicker layers produce the more commonly seen reds and greens; the thinner, more fragile tiers reveal the rarer blues and violets. The rainbow of hues exhibited in ammolite are the result of minerals combining with the original mother of pearl nacre. "The color is created by interference patterns within the crystal structure of the material," offered Issa.

Aside from aragonite, the shell preservation process may be infused with calcite, silica, pyrite, or other elements. And for spice, the conch itself may contain a number of trace elements such as aluminum, barium, chromium, copper, iron, magnesium, manganese, strontium, titanium, and vanadium. Its crystal structure is orthorhombic—so a three-dimensional rectangle of sorts. It registers at 4.5 to 5.5 on the Mohs scale of mineral hardness. Similar to the fairytale *Goldilocks and the Three Bears*, ammolite is neither too hard nor too soft; it is just right. "About 90 percent of what we find are fragments and broken pieces—that's what we turn into gemstones," shared Issa. "It's the same lapidary process as Lightening Ridge black opal."

Ammolite was registered as a gemstone in 1981 and has a price in the region of $50 per carat. It is in high demand in Asia, where it is used in feng shui, while jewelry and shaped cabochons are desired throughout the world. First Nation people are involved in the mining operation, and the tribe receives royalties based on the volume of land mined that season. "Ammolite has given the Blackfoot people a better life," declared Knowlton, one of around 1 million First Nation people across Canada.

## We Dig Iridescent Ammonites: South Dakota

South Dakota is referred to as "the land of infinite variety," in part because the scenic state is filled with fascinating geologic formations and spectacular Cretaceous fossils. The land of infinite variety is a paleontological snapshot of western North America around 100 million years ago. And it certainly was different. At that time, the western interior of the United States and Canada was traversed by an intercontinental seaway that stretched from the Gulf of Mexico to the Arctic Ocean. Some 15 million years later, the Rocky Mountains sprouted, and the waters retreated. During this process, some magnificent ammonites were preserved at the Fox Hills Formation.

"The diversity in the WIS [Western Interior Seaway] was most likely controlled by low-oxygen, bottom-water conditions; brackish surface waters derived from surrounding landmasses and the Arctic Ocean; high-sedimentation rates in nearshore settings; and substrate conditions," noted Joshua Slattery of the department of physics at the University of North Florida. "Bottom-water dysoxia and anoxia were probably the most significant influence over the fauna in deeper parts of the WIS, reflecting reduced water-column mixing."

## The Western Interior Seaway

Full of unique layers and endemic atolls, the Fox Hills preservation is a marine sandstone with shale pockets in matrix ranging in color from yellow to gray. This formation accumulated during the Late Cretaceous Period as the Western Interior waters parted, and barrier islands enlarged to become a united landmass. The vestiges are conserved as marine layers dotted with concretions filled with blue-ribbon-quality fossils. Upon hitting a sweet spot, one finds vast numbers of a few species of ammonites. Look for *Hoploscaphites nicolletii*—a curious species with an involute asymmetrical shell that tends to be shiny and smooth, with the final whorl projecting forward like a jaw with an underbite. These preservations often have an opalescent mother-of-pearl exterior. "It's a result of the preservation in sandy, mudstone concretions that allowed for a perfect quick preservation of the shell, jaws, and some organic material," explained Neal Larson.

*Hoploscaphites nicolletii* **(Morton, 1842)**
Order: Ammonitida. Appearance and lifespan: Late Cretaceous, 72.1 to 66 MYA. Location: Trail City Member, Fox Hills Formation, South Dakota, United States. Size: 6 cm.

"The fascinating thing about Fox Hills ammonites is that I never know what new discoveries we may find next. I am passionate about healed pathologies as well as diagnosing bites. Also, the quality of some of the preserved shell is the finest in the world."

Rambling through the northwestern Great Plains from Alberta, Canada, down to Colorado in the United States, the Fox Hills Formation has pockets that are a Cretaceous cephalopod utopia. Other outcrops are a dinosaur delight, with some areas hiding tyrannosaurs, while others yield marine mosasaurs.

## Along the Long and Winding Road

A satisfying fossil-hunting adventure can be had along the Native American Scenic Byway, also known as South Dakota Highway 20. It traverses the northern part of the state, running from Michigan to Montana, but the paleontological area of interest is the 250 kilometers (155 miles) between Reva Gap and the Missouri River at Mobridge. In addition to the Fox Hills Formation, you'll have the opportunity to experience the Ludlow Formation, Cannonball Formation, Tongue River Formation, and Hell Creek Formation—the last of which is the oldest Cretaceous preservation in the United States. You can also take in the Badlands and the Black Hills, not to mention the impressive and controversial human-made creation, Mount Rushmore.

*Jeletzkytes nebrascensis* **(Owen, 1852)**
Order: Ammonitida. Appearance and lifespan: Late Cretaceous, 70.6 to 66 MYA. Location: Trail City Member, Fox Hills Formation, north central South Dakota, United States. Size: 9 cm.

Figure out the whys and wherefores of digging situations by reaching out to local rockhounds. Browse the fossil forums for local preservation hunters, prospecting locations, and area collecting

## The Legalities of Fossil Prospecting in South Dakota

Public fossil hunting is allowed in South Dakota, where it is possible to find preservations on an exposed rock face, roadside outcrop, or just lying on the ground. That being said, if you find a fossil on municipal or tribal land, it belongs either to the state of South Dakota or the United States. Accessing private property requires the owner's permission.

details (like how best to transport preservations that need to be stabilized). Those who live and regularly prospect around the rivers, hills, and plains that are of interest can make great guides. They'll often escort you to their favorite collecting spots if it is a one-time experience. In the right locations, you may pick up a concretion from a stream bed, split it with a hammer, and reveal a spectacular ammonite that hasn't been seen for more than 66 million years! Digging with fossil friends who know the area tends to be far more successful than following a map that may say, "Turn right on the dirt road by the chokecherry tree and take it for 12 miles until you get to some rusty metal gates, then head another 10 miles down this road less traveled."

Once you get into the riverbed dirt, you're likely to find the number-9-shaped heteromorph *Discoscaphites conradi*, one of the species of ammonites that predominate the Fox Hills Formation. A scaphite, it has a knobby involute shell with three lines of nodes along the venter. Also look for the smooth and involute *Sphenodiscus*—this large, dynamically designed ammonite was one of the last round iterations before the entire subclass went extinct. "I have collected tens of thousands of ammonites in my life and have prepared well more than fifteen thousand," revealed Larson, the go-to specialist for extinct Fox Hills cephalopods. "It is mostly through collecting and preparation that one recognizes new and wonderful things."

**Ammonite Bite:** When prepping fossils, take the time to figure out your chunk of rock and the toughness of the preservation. Fox Hills finds tend to have a particularly hard matrix, while the opalescent ammonite hiding inside the concretion is very delicate. Patience is required.

The ammonite's opalescent nacre leads scientists to believe that these cephalopods had good eyesight.

*Cleoniceras* **Parona & Bonarelli, 1897** (sp)
Order: Ammonitida. Appearance and lifespan: Early Cretaceous, 113 to 100.5 MYA. Location: Mitsinjo, Boeny, Madagascar. Size: 12 cm.

## We Visit Ammonites: Cosmopolitan Iridescence

North America has some of the most extraordinarily well-preserved ammonites with exceptional lustrous color patterns, but it is not the only continent boasting iridescent cephalopods. Dazzling specimens have been collected from strategic fossil-gathering locations throughout the world. Some 110 million years ago off the east coast of Africa, the forests of geologically complex Madagascar became the preservation spot for a different conservation of luminous colorful ammonites. Genera like *Cleoniceras* have a rich depth of rainbow colors, robust shells, and some unique and aesthetic preservations. These fossils are desired for decorative display in addition to their paleontological attraction.

The plate tectonics associated with the breakup of the Pangean landmass resulted in a virtual latitudinal line of iridescent ammonites.

Image credit: Peter Fitzgerald, Stefan Ertmann, Joelf, Historicair, Piet-c, ARR8, CC BY-SA 4.0, via Wikimedia Commons.

We talk about climate change today as if it were something new, hah! Earth's conditions have been in flux since the planet formed some 4.54 billion years ago. At the dawn of the Jurassic, the future Europe was located considerably closer to the equator than it is now. An archipelago of islands in the warm and tropical Tethys Ocean, the area was a liquid amalgam of shallow seas, salty lagoons, marshes, and freshwater lakes. Only areas of present-day France, Spain, and Scotland broke sea level.

Europe's low-lying subtropical archipelago was a series of arid isles landscaped with low shrubs like stout woody cycads. Only small terrestrial creatures inhabited these spaces. In the air was the region's favorite dinosaur / bird, *Archaeopteryx.* There was also the *Compsognathus*, a carnivorous, bipedal theropod that grew to around the size of a turkey. Ammonites of all types frolicked in the seas and sometimes died en masse.

## The Jurassic Line

Of the many exceptional European ammonite locations are three notable European sites that produce spectacularly preserved opalescent ammonites—and oddly, they all lie along a similar latitudinal plane. During the Jurassic period, great swaths of Europe and Asia were no more than a chain of islands. Much of what is now dry land was covered by shallow seas, salty lagoons, freshwater lakes, and lush marshes. The higher elevations poked out, creating a chain of desert islands. The conditions did not suit many large dinosaurs but did favor smaller avian creatures, some of which had the capability of flight.

This period saw the fragmentation of the Pangean landmass into what we now recognize as the modern continents. Driven by plate tectonics, the North American and European landforms were split. This spread the seafloor basin, allowing for the introduction of the Atlantic Ocean, which filled the gap between the emerging regions. These conditions led to a virtual latitudinal line of conservations with spectacular ammonites.

*Psiloceras planorbis* **(Sowerby, 1824)** Order: Ammonitida. Appearance and lifespan: Early Jurassic, 201.3 to 196.5 MYA. Location: Blue Lias Formation, Somerset, England, United Kingdom. Size: 5 cm.

## We Dig Iridescent Ammonites: Across Europe

This adventure is definitely easier as the starling flies. The European sparkling ammonite route has one endpoint on an island and the other in Russia, traversing across Europe along the way. With so many countries, so many different rules, so many distinct types of preservation.

## Somerset, England

The United Kingdom is the easiest starting point for this adventure. Looking very much like swirled oil slicks, the smooth-shelled evolute *Psiloceras* is an iridescent preservation variation found around Watchet, Somerset, England. This humble compressed evolute genus is the common ancestor to all ammonite families found along the Jurassic Coast. That's because virtually all previous ammonoids faded during the Triassic–Jurassic extinction. When the prolific superfamily *Psiloceratoidea*

arrived some 205.6 million years ago, it radiated into a staggering array of niche-filling ammonite species.

This parent of area ammonites may be found compacted amid the limestone and shale layers of the Blue Lias Formation. Expect to find *Psiloceras planorbis* near the base as part of the species dominant *Planorbis* Zone. A global doppelganger, *Psiloceras planorbis* has been used to date the strata in Argentina, Austria, Canada, China, France, Germany, New Zealand, Spain, the United Kingdom, and the United States.

**Ammonite Bite:** The International Commission on Stratigraphy (ICS) has assigned the First Appearance of the Psiloceras as the defining biological marker for the start of the Jurassic Period, some 201.3 million years ago.

(*left*) ***Kosmoceras*** **Waagen, 1869**
(*right*) ***Quenstedtoceras lamberti*** **Sowerby, 1819**
Order: Ammonitida. Appearance and lifespan: Middle Jurassic, 165.3 to 161.5 MYA. Location: Łuków, Poland.
Size: largest is 5 cm.

## Łuków, Poland

Some 1,900 kilometers (1,181 miles) due east of Somerset is an iridescent ammonite niche near Łuków, Poland. Here, the chalky rainbow preservation features the prolific and always popular *Kosmoceras* (also known as *Cosmoceras*). Younger than the Somerset preservation, this cluster in eastern Poland dates from the Middle Jurassic, about 165 million years ago. The natural mother-of-pearl shell has been preserved in such a way that prepped specimens sparkle like opals. Be advised, unless they are heavily lacquered, the oxygen in the air may decompose the pyrite-infused preservation into powder. The process depends on climactic conditions.

*Kosmoceras* is a fine-looking European ammonite genus that radiated into more than one hundred species. It can be identified by its moderately evolute shell with irregular ribbing and a simple aperture. There may be a row or several of lateral knots, while knobs bracket and protect a low-lying venter.

## Volga River, Russia

Head further east from Łuków, pass behind the Russian border, through the Ural Mountains, and just keep going. When the odometer indicates that you have traveled around 2,000 kilometers (1,243 miles) east of Łuków, you will have arrived at the Volga River location. A memorable adventure, the area yields magnificent ammonite specimens. from iridescent rose to ruby red, locally known as "fire opal." The lack of acidity in the depositional environment preserved the blushing shimmer. The ideal conservation situation enhanced the luminous rose rainbow color in the shells of Jurassic species such as *Peltoceras* and *Kosmoceras*.

*Peltoceras* **Waagen, 1871**
Order: Ammonitida. Appearance and lifespan: Middle Jurassic, 166.1 to 163.5 MYA. Location: Odintsovo, Moskovskaya, Russia. Size: 4.5 cm.

Curiously, the Volga River area went through two phases of iridescent fossilization. Also found in the area is the Early Cretaceous *Craspedodiscus*. An offshoot from the ammonite superfamily *Perisphinctaceae*, this involute flying-saucer-shaped genus lived about 133 to about 129 million years ago and was one of the boreal species endemic to Northern Europe and Siberia. "Much of ammonite fossilization depends on being entombed in water-resistant mud that turns into shale. High-latitude fossils fare better than those from wet humid areas," observed Neil Landman of the American Museum of Natural History. "The worst preservation usually is from sandy areas."

## The Legalities of Fossil Prospecting in Russia

We've already reviewed the rules about taking invertebrate fossils home from the United Kingdom and Poland, so now let's turn to Russia. If you are bringing natural history items of value out of the country, export licenses are required and are issued by the Ministry of Culture. These documents are often countersigned by directors of the Institute for Paleontology in Moscow. This wing of the Russian Academy of Sciences advises the ministry about whether the fossils are scientifically important and thus ineligible for export.

Make life easy on yourself. Get a letter from the Russian Academy of Sciences to put with your specimens . . . or just buy them online.

*Quenstedtoceras lamberti* Sowerby, 1819 (sp)
Order: Ammonitida. Appearance and lifespan: Middle Jurassic, 164.7 to 161.2 MYA. Location: Dubki locality, Saratov Oblast, Russia. Size: largest is 8 cm.

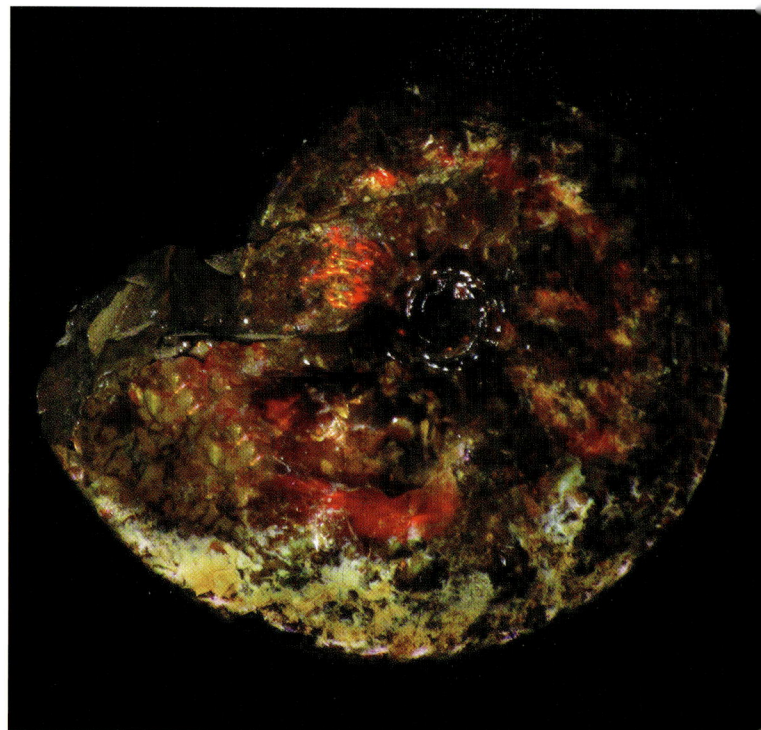

*Placenticeras* Meek, 1876
Order: Ammonitida. Appearance and lifespan: Late Cretaceous, 83.5 to 70.6 MYA. Location: Bearpaw Shale, Saint Mary River, Alberta, Canada. Size: 11 cm.

## More Luster

Ammonites of extraordinary preservation and luminous color are regularly uncovered in new locales. They may be one-off finds, like a rare *Protexanites* in Japan, *Libycoceras* in Nigeria, *Submortoniceras* in South Africa, or the *Calliphylloceras* in India. Additional special preservation locations for ammonite fossils with subtle color and iridescence have been recorded in Normandy, France; Coon Creek, Tennessee; Mississippi; California; and Utah in the United States; Australia; Lebanon; and Morocco, among other locales.

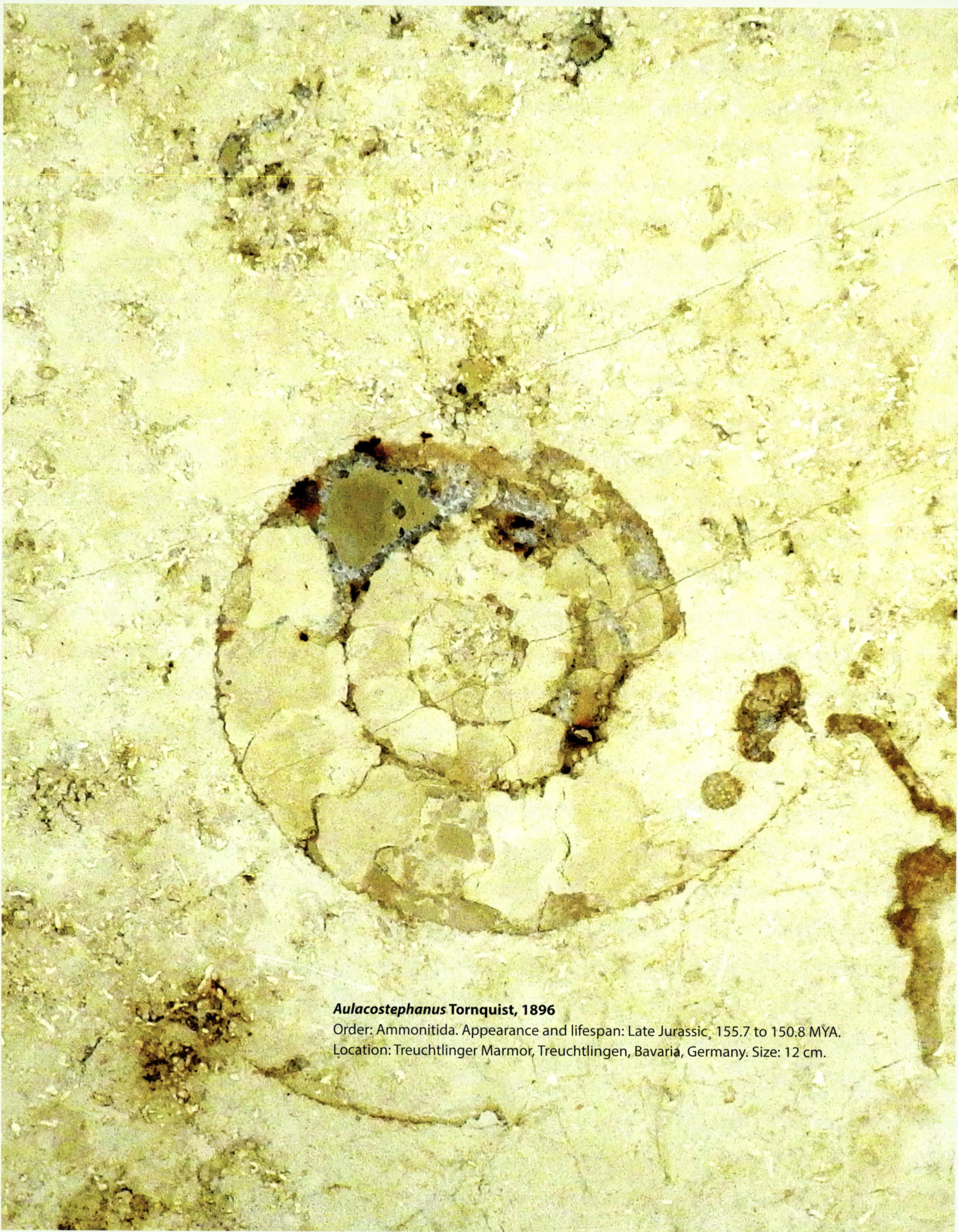

**Aulacostephanus** Tornquist, 1896
Order: Ammonitida. Appearance and lifespan: Late Jurassic, 155.7 to 150.8 MYA.
Location: Treuchtlinger Marmor, Treuchtlingen, Bavaria, Germany. Size: 12 cm.

# The Animal in the Ammonite Shell

**M**ost fossil enthusiasts fantasize about going on a dig and finding a specimen that holds the possibilities of expanding the boundaries of science. In the cephalopod world, one of those great magical moments occurred in 1966. Just as Beatlemania was rocking the Western world, Mackenzie Gordon Jr. and Gerhard Ludwig Closs became the first scientists to record information about preserved ammonoid soft-part preservation. Gordon was an accomplished invertebrate paleontologist with the United States Geological Survey (USGS), and Closs was with the United States National Academy of Sciences. They described a phosphatized Carboniferous goniatite named *Glaphyrites* (also known as *Eoasianites*), first discovered and described in Uruguay. They were also the first to document an ammonoid's digestive system, and their efforts added a new chapter to the fossil enthusiasts' book of famous finds.

Prior to twenty-first-century breakthroughs, ammonoids were a peculiar paradox. Pretty with a purpose, their decorative spiral shells are some of the most iconic fossils ever unearthed, yet we are just now beginning to figure out the nonbiomineralized creature. New technologies allow science to recognize and extrapolate on unique remnants from the distant past. Molecular microscopes and customized lighting allow for more intensive study, with researchers now able to identify a myriad of previously unknown or unseen morphological features, including the hooks at the end of ammonoid arms, and confirming the arm crown. Notable soft-body material from Solnhofen has also been

***Glaphyrites* (Ruzhencev, 1936)**
Order: Goniatitida. Appearance and lifespan: Carboniferous, 326 to 279 MYA. Location: Pumpkin Creek Formation, Daube Ranch, near Ardmore, Oklahoma, United States. Size: 3 cm.

described. Teams of fossil researchers under the leadership of Christian Klug of the University of Zürich, in Switzerland, used Solnhofen ammonite fossils to identify the organs and body parts of the creature inside the shell. As a key component of their investigation, they photographed the specimens under both white and ultraviolet light. "The latter helps to make faintly phosphatized structures better visible," noted the team.

## Class Cephalopod

All ammonoids are cephalopods—a class of predatory marine molluscs that adapted their body plan to a more active and predatory lifestyle than its benthic predecessors. Cephalopods are characterized by a prominent head, bilateral body symmetry, and their ability to swim. This group transformed the slimy muscular foot molluscs used to stick to rocks and slink along the ocean bottom into a hydrodynamic torso that allowed cephalopods to propel themselves through the seas. This ancient class of creature had a sophisticated eating apparatus. A crown—typically of eight or ten appendages—grew around the cephalopod's mouth allowing, it to pull food aggressively into its maw. This grouping of muscular hydrostats—also known as arms or tentacles—was a set of complimentary tools—and another defining class characteristic.

There have been four cephalopod subclasses throughout paleo history, and three have shells. Ammonoids and orthocerids are extinct; the nautilus struggles to continue its 500-million-year-or-so reign of survival in our planet's warm deep oceans. Alive and flourishing are the *Coleoidea*. These distant relatives include octopuses, squid, and cuttlefish—all of which have their protection on the inside, and many of which possess the ability to squirt an inky diversion so they can propel off to safety. (Ammonoids do not have this ability.)

Ammonite eyes would have been similar to those of the chambered nautilus.

**Nautilus pompilius Linnaeus, 1758**
Order: Nautilida. Appearance and lifespan: Miocene, 11.6 to 5.3 MYA Location: Nautili are endemic to the Indo-Pacific Ocean area, where they inhabit the deep slopes of coral reefs. Size: approximately 16 cm.

An example from each of seven extant orders of cephalopods. Clockwise from top left: **Loligo vulgaris**—common squid, **Eledone moschata**—octopus, **Nautilus pompilius**—chambered nautilus, **Vampyroteuthis infernalis**—vampire squid, **Austrorossia mastigophora**—bobtail squid, Spirula spirula— deep water squid, **Sepia officinalis**—cuttlefish.

While sophisticated life above the waterline has iron-based red blood coursing through its system, today's cephalopods have copper-based anatomical fluids. As a result, the stuff that flows through their bodies is colorless until it is exposed to air, where it turns into a translucent teal. Malacologists have extrapolated that ammonoids had blue blood like today's cephalopods.

Ammonite eyes should have functioned like those of most modern cephalopods. Like the nautiloid, they were large and on the side of the animal's head, behind its arms. Although not as complex as the human eye, their variable and iridescent shells confirm that this subclass could see well enough. The creature could sense vibration but didn't have any hearing. Paleozoic ammonoids had chitinous jaws, similar to modern coleoids. By the Mesozoic, these spiraled cephalopods fronted an aptychus beak and jaws, made up of proteins with a calcite exterior for added strength. Inside their orifice, as part of their mouth apparatus, these ammonoids sported a raspy radula for shredding prey. Their siphuncle managed buoyancy, and the hyponome supplied propulsion.

## a, reconstructed anatomy

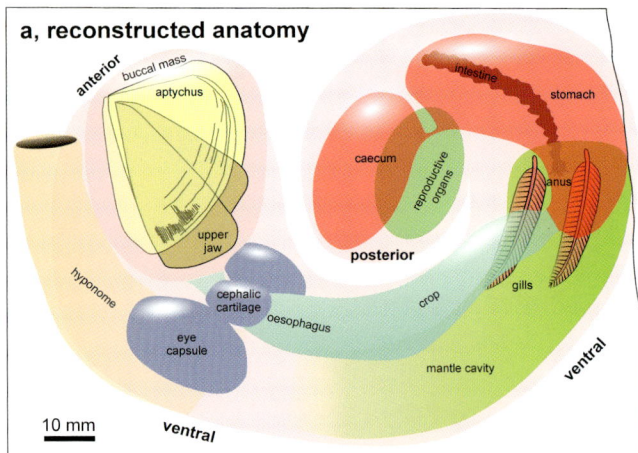

## b, anatomy in the conch

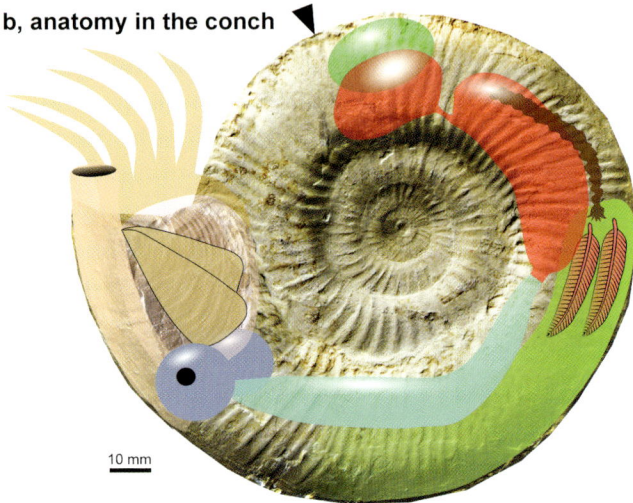

Reconstruction of an ammonite's soft body parts based on a preservation found in Solnhofen.

Image credit: Courtesy of Dr. Christian Klug.

## The Science of *Jurassic Park*

Because there has always been a dearth of available nonbiomineralized ammonoid material for scientific research, scientists have extrapolated about these extinct sea creatures using observational studies of similar living cephalopods. Thanks to refined detection processes, we have reached the time when reconstructions of long-extinct species are now a common way of study. Not long ago, a team from United Kingdom Research and Innovation used their neutron and muon instruments to examine a Jurassic ammonite fossil on an atomic

scale. "We found evidence for muscles that are not present in Nautilus, which provided important new insights into the anatomy and functional morphology of ammonites," observed lead author Lesley Cherns, honorary research fellow from Cardiff University's School of Earth and Environmental Sciences. Among their conclusions were that the muscles that attached the animal to the shell allowed the creature to lengthen and contract, which left connection scars near the rear and sides of the living chamber, adjacent to the largest septum (where the siphuncle would adhere). The study confirmed that ammonoids avoided predation by retracting into their shells.

## Be the Ammonoid

Many who envision an ammonite imagine the dazzling preservation of their coiled chambered home. But the animal that inhabited those chambers is also a vital part of this cephalopod's story. With that in mind, let's pique the intellectual curiosity of sea creature fans with this simple and entertaining hand puppet that artfully explains the appearance of the animal in the ammonite shell. Please wear something with long sleeves to allow the cuff area to masquerade as the conch.

1. Pretend your right hand is the ammonite animal, and the long sleeve is the shell. With your hand flat and thumb down draw your hand up into your sleeve; this is how an ammonite retracts into its conch. Your extended fingers represent the arms that grasp prey. Stretch your thumb toward the bottom sleeve margin and think of it as the hyponome—the funnel through which jets of water pushed the ammonite backward through the seas.

2. The ammonite's mouth was located where your major knuckles sit, at the base of your fingers. Imagine it is surrounded by a horny jaw

*Nautilus pompilius* is the best-known extant species of nautilus. The shell, when cut away, reveals a lining of lustrous nacre and displays a nearly perfect equiangular spiral. The counter-shaded shell is a source of protection. When seen from above, its dark top blends in with the darkness of the sea. From below, its white bottom blends in the light from above.

**Nautilus pompilius Linnaeus, 1758**
order: Nautilida. Appearance and lifespan: Miocene—11.6 MYA.
Location: Western Pacific and Indian Oceans. Size: up to 25 cm.

Image credit: © Hans Hillewaert.

Ammonite with aptychus in situ.

**Lithacoceras Hyatt, 1900** (sp)
Order: Ammonitida. Appearance and lifespan: Late Jurassic, 157 to 145 MYA. Location: Solnhofen Limestone, Eichstatt, Germany. Size: 15 cm.

resembling a beak. Known as the aptychus, it allowed ammonites to crush prey.

3. Behind its mouth were the eyes. The left optic orb would be found just beyond your first knuckle.

## *A* Is for Aptychus

A curious piece of paleontology, the aptychus was named long before someone figured out they were ammonoid jaws. It's primary purpose was as the first step in the process of devouring prey. "All ammonoids had jaws," confirmed Christian Klug. "The presence of these beaks/jaws in several Devonian lineages represent good evidence for this."

When it comes to fossilization of the soft-bodied animal, the physical attribute that preserved most regularly was the aptychus. Serving as the ammonoid's upper and lower jaws, these curved shelly plates covered the aperture, similar to a mollusc's operculum. They varied in number, size, shape, and surface structure, sometimes resembling two halves of a paper heart. Throughout the fossil record, notable numbers of ammonoid aptychi have been found close to the apertures of fossil conchs. Analysis of shelled cephalopods in situ has allowed paleontologists to confirm that the aptychus was, among other things, a powerful adaptation of the jaw. Finding an ammonite plate with aptychi is always a prize. These are not that

*Lithaceras* and *Subplanites* with aptychus.

***Lithaceras* Hyatt, 1900** (sp)
Order: Ammonitida. Appearance and lifespan: Late Jurassic, 157 to 145 MYA. Location: Solnhofen Limestone, Eichstatt, Germany. Size: 15 cm.

uncommon among the finds in southern Germany's Solnhofen Limestone, a Jurassic Konservat-Lagerstätte. "Sometimes you see patches of organic matter and you don't know what it is. But organic parts which are identifiable (i.e., which organ) are extremely rare, except for chitinous parts such as nonmineralized jaws," remarked Klug. "They can be quite common in some strata and localities."

## Solnhofen Is So Rich

Some 150 million years ago, southern Germany was part of an archipelago in the shallow warm tropical Tethys Sea. For around 5 million years in the Late Jurassic, this tranquil, temperate archipelago with lush blue lagoons was a paradise for fauna and flora—until it wasn't. During its sea-to-land transition, the area we now know as Europe became increasingly less accessible to the open ocean. The mire became a sort of quicksand, and wildlife became entombed in the lime slurry at the bottom of these inlets. Organisms that fell, drifted, or got washed to the briny bottom were enshrouded in soft, sandy, blonde carbonate mud. As the lagoons dried out, the mire developed into slender layers of fine-grained limestones interstriated with shale. Now known as the Altmühl Valley, it is the location of the Solnhofen Limestone. The formation is part of a vast deposit some 15 kilometers (9.3 miles) wide and 150 kilometers (93 miles) long that rambles through southwestern Germany's Jura Uplands. This conservation has been extensively quarried since the Stone Age, when the easily cleaved limestone was used for roofing and floor tiles.

Known regionally as *plattenkalk*, the utility of the limestone preservation, expanded in the late eighteenth century. Author and actor Alois Senefelder devised the printmaking process of lithography,

**Ammonite Bite:** Mining has been integral to the fossil discovery process. Many of the conservations exhibited in museums today were originally found by mineral prospectors and quarry workers.

etching into flat slabs of Solnhofen limestone to print theatrical works. This cost-efficient planographic process is still in use today.

The true paleontological value of this area of Bavaria was not realized until well into the nineteenth century, when local scientists and officials began curating the treasures extracted from the Solnhofen deposits. It was then realized that the lithographic limestone was laden with all sorts of special preservations. The rapid burial of creatures in the powdery sediments of these stagnant marine basins has created fossils with exquisite detail. Intricacies include fragile or soft-bodied organisms and other late Jurassic rarities seldom preserved, which is why Solnhofen has been awarded Konservat-Lagerstätte status. "Bear in mind that there are tens of quarries with quite variable preservation/taphonomic conditions around Solnhofen," explained Klug. "Shallow basins between reefs often provided quiet water conditions, locally combined with microbial mats and/or poor oxygenation."

There are several commercial prospecting locations around Mörnsheim, Blumberg, and Solnhofen where rockhounds can discover fossils for the price of a few euros. The chance of being the recreational paleontologist to uncover the big game of this Lagerstätte brings curious rockhounds to the Altmühltal Valley. The Eichstätt railway station in the district capital is an excellent place to begin this fossil journey. It's an easy two-hour train ride going north from Munich.

The station is the start (and finish) of an entertaining loop trail that offers a fine orientation to the paleontology of the area. Known as the Fosilienpfad (Fossils Way), this 9-kilometer (5.6-mile) loop passes several of the valley's major sites and fossil collections. During your exploration, you will notice that numerous commercial excavations in the area supply stone for lithography and architectural uses. Most are closed to the random fossil hunter, but there are legal and inexpensive fossil excavation opportunities. Along your promenade, note the location of the Museum Bergér excavation pit. When you are ready to dig, residents suggest grabbing a hat and a couple of liters of water and heading over to the institution's prospecting location. Named Steinbruch am Blumenberg, it is one of the region's most popular fossil-hunting areas because the stones are relatively soft and digging equipment can be rented on site. The Bürgermeister-Müller Museum also offers quarrying opportunities, as well as some of the finest examples of area discoveries.

Inside the field pit lies an expansive excavation peppered with rockhounds splitting stones. Some say it can be hard to find specimens, but if you spend about three hours prospecting, you will come away with treasures in your possession. (Specimens can be fragile. Have a box on hand for storage and transport.) As is true throughout the world, a large part of fossil hunting is skill. You need to get a feel for where the desirable pockets may be located. Some areas are rich with preservations, others, not so much.

There is a comprehensive cornucopia of Late Jurassic marine and terrestrial biodiversity conserved in the soft lagoon muds. The most common day-dig fossil find is the lacy starfish-like crinoid, *Saccocoma*. There are also marine coprolites—also known as *Lumbricaria*—that vaguely resemble tiny sausage strings of stone poop. Along with ammonites, regular discoveries from the plattenkalk include crustaceans, scaled fish, plants, and terrestrial vertebrates—including twenty-nine species of pterosaur. Perhaps the most famous Solnhofen fossil is *Archaeopteryx*, one of the earliest known birds that left impressions of its feathers preserved in the rock.

**Ammonite Bite:** Read the fine print on the quarry contracts. Should you find something rare, the property owner likely has the rights to the discovery.

**Perisphinctes Waagen, 1869**
Order: Ammonitida. Appearance and lifespan: Late Jurassic, 161.2 to 155.7 MYA. Location: Malm Delta, Treuchtlingen, Bavaria, Germany. Size: 20 cm.

## We Dig Ammonites: Solnhofen, Germany

Ammonite aficionados will delight in the dozen or so different genera regularly recovered from the limestone—and if you are really lucky, the aptychus may be nearby. In these subtidal carbonates, you may discover the venerable smooth-shelled *Phylloceras*, the thin-ribbed *Perisphinctidae*, and the gracefully evolute *Lithacoceras*. And you definitely want to find or buy a thin-ribbed *Subplanites*, one of the more common cephalopod genera in the formation and the genus of a famous area discovery. One particular perisphinctid ammonite that is approximately 150 million years old was preserved with various parts and organs intact. It's been suggested that the non-biomineralized animal was separated from the conch during a failed predation, only to become covered in silt. Researchers have a realistic view of the ammonite digestive tract, male reproductive organs, mantle cavity, gills, and hyponome. "The story is fascinating. We are always focused on looking at the shell to find the remains of the soft body. Who knew we should be looking elsewhere?" observed Neil Landman, invertebrate paleontologist emeritus with the American Museum of Natural History. "The soft body is pretty much what we imagined. The arms always seem, however, to prove elusive—I guess they are just flimsy material."

## Plattenkalk Prep

Many Solnhofen specimens are found by separating layers of limestone matrix. Oftentimes, pieces don't divide accurately, and details may be missing, or after more than an era of being safely hidden in the damp ground, the frail layers of rock get cracked during excavation. And once the stone comes out of the earth, it may go through a desiccation process and become dry and delicate, resulting in pieces of your find layering off. "The easy way to stabilize a stone is a thin super glue," shared expert fossil preparator Ben Cooper. "There are many options out there, so my advice is to research online, to find what suits the task. After the glue sets, it should be wrapped so that all pieces stay with the main specimen, to attach later if needed. Materials used could be plastic wrap, aluminum foil, plaster, et cetera, depending on the piece."

There are ways to expertly repair and recondition most fossils so the restoration does not detract

> **Ammonite Bite:** The Jura Museum and castle in Eichstätt shares some fine examples of the more than eight hundred identified vertebrate and invertebrate animals, land plants, and unicellular protists found in the formation.

The Jurassic Konservat-Lagerstätte of the Solnhofen Plattenkalk is renowned for its abundance of ammonite aptychi, such as this *Lithacoceras* with anaptychus.

*Lithacoceras* **Hyatt, 1900** (sp)
Order: Ammonitida. Appearance and lifespan: Late Jurassic, 150.7 to 145.5 MYA. Location: Solnhofen Limestone, Eichstatt, Germany. Size: 4.5 cm.

Octopuses have eight flexible arms covered in suckers. These appendages give this most sophisticated cephalopod the ability to saunter, grasp objects, and control its environment.

Image credit: Pseudopanax at English Wikipedia, public domain, via Wikimedia Commons.

from the beauty of the specimen. Solnhofen prep experts use a tint and brush to touch-in missing details and complete pieces. During the summer, the Günzenhausen Archeology Museum offers a specialty workshop in fossil restoration, reservations are necessary. This jewel of a regional museum offers the historical cultural landscape of Franconia.

If you are unable to unearth the piece of your dreams, there is the option of purchasing your memories. Fossil emporiums dot the Altmühltal Valley. The gift store at the Museum Bergér is a fine place to start or end your shopping expedition. Keep your expense receipts with your specimens in case your customs agent is interested.

## Arms or Tentacles?

Ammonites had arms; nautiloids had tentacles. What's the difference? Tentacles are generally

**Ammonite Bite:** The world's most intelligent living invertebrate, the octopus has eight sucker-covered arms. Octopuses favor their first three pairs of arms for grabbing and handling objects. Those in cephalopod circles adopt an anthropomorphic philosophy, acknowledging that octopuses have six arms and two legs.

longer and have hooks or suckers only near the end of the appendage. Arms, like those on an octopus, feature suction cups racing down the entire length of the limb. Some cephalopods have arms, some have tentacles, and some have both. Interestingly, not all tentacles are alike, giving the creatures an array of utensils at their disposal.

Squid and cuttlefish are known as 10-armed cephalopods; they have eight short arms and two

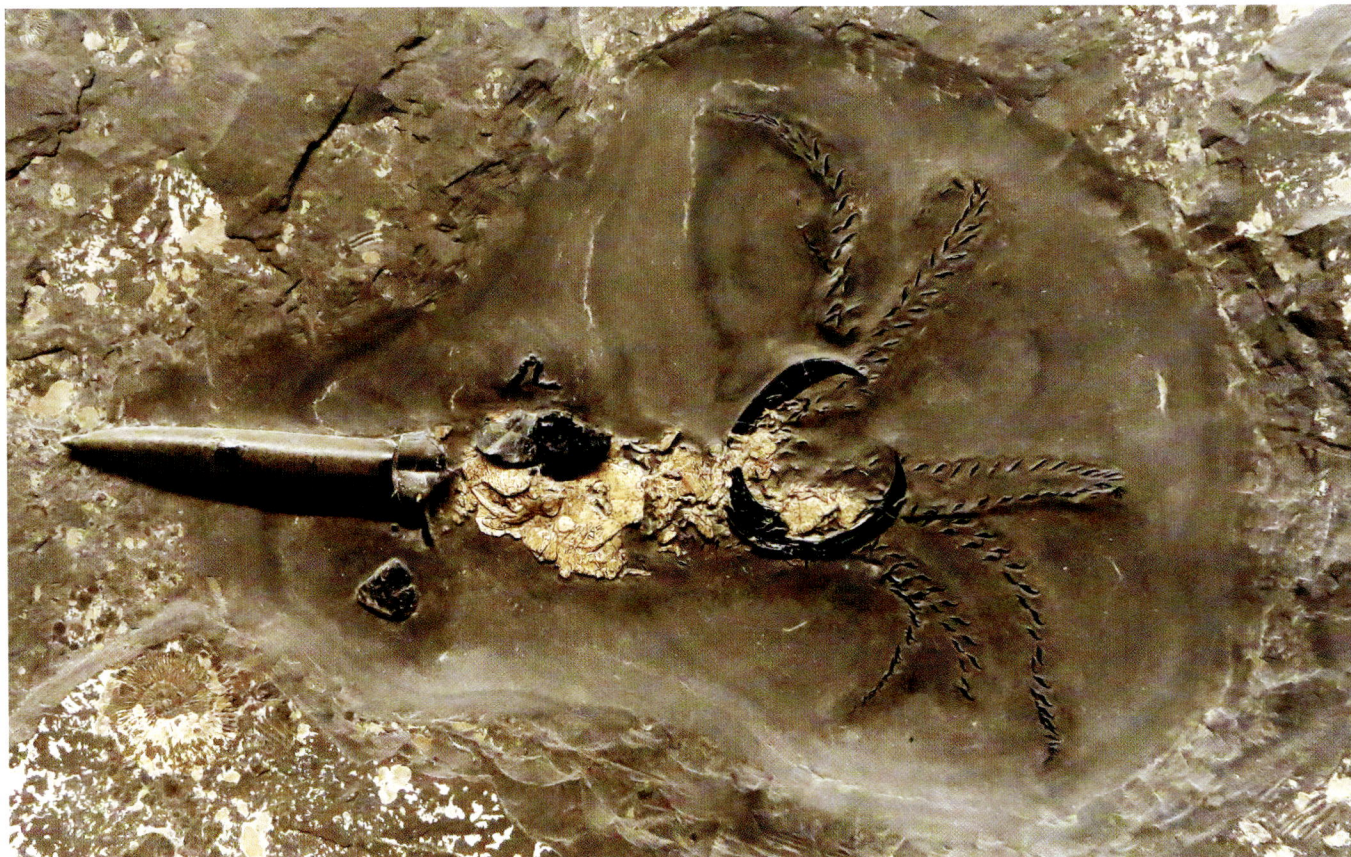

An incredible preservation of a belemnite with soft body parts. These Jurassic coleoid cephalopods typically had around forty hooklets on each of their ten arms as well as a large pair of menacing chitinous hooks known as onychites.

**Passaloteuthis bisulcata (de Blainville,1827)**
Order: Belemnitida. Appearance and lifespan: Early Jurassic, 183.7 to 175.6 MYA. Location: On display at the Museum am Lowentor, Stuttgart, Germany. Size: Body length median was approximately 100 cm, arm length median was 25 cm.

long feeding tentacles. The two special appendages shoot out like harpoons and grab unsuspecting prey quicker than the blink of an eye; often, their chow never saw it coming. If you haven't seen these vicious hunters in action, check out videos on squids feeding. They are simultaneously shocking and entertaining.

Arm development in cephalopod embryos begins with ten original buds for both squid and nautili. Extensive investigation has surmised that *Ammonoidea* likely had ten tentacles situated around its aptychus. When it comes to fossilization, if these appendages are preserved at all, they typically look like strategically located tar-colored strands. Laypeople might think it's some disparity in the rock, but there are amazing paleontologists who spend their careers with electron microscopes interpreting markings such as these. Detailed analysis has allowed the world's most esteemed ammonite authorities to adopt the crown of arms death theory as part of the life and habits of this extinct cephalopod.

## The Arm Crown

Those who have watched a nautilus in its death throes note that the animal retracts into its shell to die. Its many tentacles are drawn in with the tips joined outside the aperture, rather resembling a crown. Applying cephalopod logic, it's likely that, when threatened, the ammonoid animal instinctively pulled back into the living chamber of its conch, allowing only a tiara of tips to dangle out of its orifice. Using this working premise, researchers studied fossilized ammonite marks and blotches from several Konservat Lagerstätten locations along what was North America's Western Interior Seaway. They hit a motherload of new information. Revealing fossil finds offering soft-bodied information were discovered in some *Scaphites* found at the Fox Hills Formation of South Dakota and the Bearpaw Shale of Montana. Paleontologists uncovered nearly fifty finely preserved Late Cretaceous ammonoids with a series of fossilized hooks around the aperture. These indicators confirm that the ammonites' arms had retracted into the body chamber as part of the death process. These Western Interior Seaway specimens display a termination crown. The trained eye and microscope also revealed two rows of paired hooks. In-situ jaws are part of this prehistoric cephalopod bonanza, indicating that the non-biomineralized animal was still inside the shell during fossilization, even though it was not preserved. "Presumably, the arms would have retracted into the body chamber directly preceding (due to stress) or following the death of the animal," concluded Landman, one of the authors of the study. In Europe, a team from the University of Zürich (under the auspices of Klug) discovered another small pocket of Late Cretaceous ammonoid arms in the Hesseltal section of Germany's Söhlde Formation.

> **Ammonite Bite:** Some hooks occur below the jaws, suggesting that the creature had at least one ventral arm pair, a biology that is still present in some modern cephalopods.

## We Dig Soft Body Parts: The Hesseltal Formation, Germany

Not much happens in the bucolic village of Lengerich, a peaceful burgh of about 23,000 residents north of Germany's Central Jura Uplands. People tend to be in a laid-back holiday mood as they enjoy the traditional Teutonic ambiance of this historic hamlet. Visitors might spend the morning peering across fields with binoculars appreciating the bird life flitting among the reeds. In the afternoons, some sightseers canvass the fresh meadows on horseback or bicycle, while others hike upon leafy floors of the deep green deciduous forests. Paleontology fans in search of the next great discovery will survey the verdant Teutoburger Wald and notice a massif of three hilly ridges. Their fossil-finding instincts perk up as they eye the outcrops of finely layered sediments, hiding stories from a variety of geologic events. The choice is the gray flaky marls of the Hesseltal Lagerstätte, which are reputed to hide Late Cretaceous cephalopod soft-body parts.

Like finding a golden ticket in a chocolate bar, these marlstones have yielded seventeen sediment-compacted baculitid ammonoids with carbonized and partially phosphatized soft parts. The elite European team that described these specimens was able to confirm previously unrecorded information about *Ammonoidea*. Digestive discoveries included aptychi (occasionally articulated) radulae, and some food bits, which provided insights into the cephalopod's diverse diet. Also identified were parts of the brain casing, digestive system details, and siphon tube, as well as a sampling of internal organs. At least one of these special baculitids were preserved with a crown of ten short and skinny retractable arms surrounding its mouth. A surprise was the discovery of striated color patterns, which are still under investigation and interpretation.

This **Cleoniceras** ammonite fossil has a well-preserved siphuncle along its ventral margin.

*Cleoniceras* **Parona & Bonarelli, 1897** (sp) split.
Order: Ammonitida. Appearance and lifespan: Early Cretaceous, 113 to 100.5 MYA. Location: Mahajanga, Madagascar. Size: 8 cm.

## The Siphon's Uncle?

Reaching through the interior of the magnificent ammonite shell is a tubular structure called the siphuncle. For approximately 500 million years, shelled cephalopods have used some variant of this viscous tube and duct structure to regulate its buoyancy. A significant natural gift, the siphuncle regulated the volume of liquid versus gas in each of the ammonoids' shell chambers. The ability to absorb and expel liquid allowed this cephalopod to navigate the water column. The siphuncle design might be thought of as a rustic version of an elevator shaft in a tall building. Consider each septum to be a floor that was serviced by a membranous connecting tube—in other words, the siphuncle. "It is believed that, by secreting gas and fluids into the hollow shell chambers through their permeable siphuncle, they could regulate the air pressure in the chambers and thus control their balance, buoyancy, and depth in the water column," explained a spokesperson from the Aquarium of the Pacific in Long Beach, California.

One of the sleekest models wading around the Early Mesozoic oceans, ammonoids were carnivorous predators, basically eating whatever tiny organisms happened to cross their path. The animals had an efficient feeding apparatus surrounded by that ominous ring of arms. Their morphology facilitated the manipulation of tiny or sluggish prey from the water into their mouths. Think of ammonoids as pocket-sized predators consuming plankton and small crustaceans. The ammonoid's feeding system is known as the buccal mass.

Instead of a tongue and teeth, ammonite prey was gnashed by a radula. This sharp, curved comb-like structure, with rows of high, slender pointy teeth, shredded and filtered the meal. Now ready to become nutrition, the morsels either passed directly into the stomach or were stored in the gullet for future use.

Living cephalopods and many extinct ammonoids usually have a hardened chitinous beak at the mouth opening, with the ribbon-like radula sitting

Ceratite with nonbiomineralized remains by the aperture.

***Spirogmoceras shastense* Smith 1904**
Order: Ceratitida. Appearance and lifespan: Late Triassic, 237 to 222 MYA. Location: Brock Mountain, Lake Shasta, California, United States. Size: 1 cm.

> **Ammonite Bite:** Researchers have identified notable differences in ammonoid mouthparts, indicating that diet varied among species.

directly behind it. Most of the tearing and crushing of food was done by the beak. The tongue-like belt of teeth of the radula further gnashed up the meal and pulled food particles down into the gullet.

One of the more obscure bits of *Ammonoidea* non-biomineralized evidence turned up at Nigh Kap Stosch, Greenland, where there has been an occasional find of ammonites preserved with curious squiggly lines near a couple of blotches. Microscopes have confirmed that, indeed, some fossils from the earliest Triassic species such as *Ophiceras greenlandicum* display something that resembles a dry fern leaf in the body chamber between remains of the stomach and the buccal mass. Feathery structures resembling branched gills have also been identified from the Triassic ceratite superfamily *Otoceratoidea*. Moreover, a similar aeration filament has been connected to an *Eleganticeras* (also known as *Cleviceras*) from the Early Jurassic of northern Germany. Undoubtedly, plenty of information has yet to be deciphered and explained. If you keep in mind that blotches and discoloration could actually be soft body parts, you will scrutinize the fossil matrix in a different light.

## Buoyancy

Ammonoids were the first scuba divers able to navigate the water's depths. To rise, this cephalopod absorbed elements from seawater to increase the saltiness of the fluid in the siphuncle. Through osmosis, the less saline liquid in the camerae was drawn into the mineral-rich siphon system. At the same time, gas—likely some combination of

nitrogen, oxygen, and carbon dioxide—diffused from the siphuncle to replace the water in the emptying chambers. With lighter elements in its cavities, the ammonoid weight decreased, thus increasing buoyancy. Now lighter, the cephalopod was able rise in the water column.

To dive, the technique was reversed. Pumping ions—typically sodium and chloride—into the camerae increased the salinity of the internal fluid. Water was then absorbed via osmosis; gasses were pushed out, density increased, and the ammonite grew heavier and dropped to deeper ocean depths as buoyancy decreased. Gas moved between the siphuncle and the shell camerae via a passive process. The cephalopod expended energy absorbing the water from the chamber. Only class members with chambered shells have siphuncles. Some coleoids, like cuttlefish, still have marginal renditions of this siphon system.

## Do the Locomotion

An ammonite is an example of Newton's third law: for every action, there is an equal and opposite reaction. Ammonoids nipped through the warm, shallow Mesozoic seas—shell first—by squirting jets of water from their bodies through a hyponome. The ammonite's propulsion system was positioned below the arm crown. This cephalopod sucked in water through the sides of its head; pumped it over their gills; then expelled a stream through that funnel-like opening at the bottom of

Goniatite split displaying the central siphuncle and backward-facing septa.

*Cancelloceras* **Ruzhencev & Bogoslovskaya, 1975** Order: Goniatitida Appearance and lifespan: Carboniferous, 318, to 315 MYA Location: Benbulben Shale Formation, County Sligo, Ireland Size: 4.5 cm

its aperture. The equal, opposing forces of drawing the water in and spitting it out sent the cephalopod scooting in the opposite direction from its squirt.

## Dobravlje, Slovenia

Some of the most interesting remains of what could be ammonite propulsion organs were found in Dobravlje, Slovenia, in 1996 by Herbert Summesberger and collaborators from the Vienna Museum of Natural History. Unlike other visitors, they were not in town to see the Skocjan Caves—a UNESCO World Heritage Site of underground drainage systems with sinkholes and caverns formed by the erosion of the area's more soluble rocks. Nor

*Prolyelliceras ulrichi* **(Knechtel, 1947)**
Order: Ammonitida. Appearance and lifespan: Early Cretaceous, 113 to 105 MYA. Location: Huanuco, Perú. Size: 9 cm.

Image credit: José Juárez Ruiz.

**Ammonite Bite:** The ammonite moved by the same principle that launches a rocket. Known as jet propulsion, the craft is pushed in the opposite direction of the detonating fuel thrusting out of its engines.

were they cycling through the pastoral landscapes and vineyards of Vipava Valley. Summesberger and colleagues got up to their knees in limestone chips, fascinated by an approximately 80-million-year-old ecosystem of compacted ammonites with preserved body chambers and aptychi partially in situ. As they chiseled, they found *Placenticeratidae* with previously unrecorded paintbrush-like structures. These mysterious forms that were near the ventral side of the aperture were later determined to be part of the hyponome. The spray of minute and battered brachiopods has been interpreted as the contents of the crop or the stomach of the decaying ammonite. This poetic preserved portrait of death was capped off by roll marks left by the ammonite's keel as it drifted to its final resting place.

## Extra Credit: Engineer a Jet Propulsion System

You, too, can create your own jet propulsion system. For this simple and silly experiment you will need a drinking straw, a length of string, a balloon, and some tape.

1. Thread a plastic drinking straw onto a length of string.
2. Tie each end of the string to sturdy anchor points that are about the same height. (You can use two doorknobs, chairs, cabinet handles, etc.)
3. Move the straw to one end of the string.
4. Blow up a balloon and pinch it shut (don't tie it).

5. Tape the balloon to the straw, with the open end of the balloon toward the nearest anchor point.
6. Let go of the pinched end of the balloon and *zoom*! The balloon straw will go zipping down the string. The air escaping the balloon provides the fuel for jet propulsion, demonstrating how an ammonite would have moved through the water.

## Mile-High Marine Creatures

Have you ever seen pictures of two cars crashing at high speed? The force of the impact twists and presses the vehicles into a craggy metal heap. Instead of automobiles, now imagine two chunks of Earth's crust being pushed into each other as the seafloor expands. Can you envision the rough, rugged, and distressed mountain belt created when pieces of Earth's top layer collide with one another? This advancing land smash built the longest continental mountain range on the planet—the Andes cordillera. The backbone of South America, this Alpen chain stretches some 7,000 kilometers (4,350 miles), providing an extraordinary example of plate tectonics and preserving marine creatures like ammonites in mountains more than a mile high in the sky.

## We Dig Ammonites: Cañón de Cotahuasi, Peru

**Landscape Reserve Sub Cuenca del Cotahuasi**
**Location: Arequipa Region, La Unión Province, Peru**
**Stratigraphic Range: Jurassic-Cretaceous**
**Date of UNESCO Submission: May 2019**

If you like heights with your ammonites, Cotahuasi Canyon in Peru may be an ideal adventure for you. Bear in mind, however, that this South American expedition is not for the hurried or the faint of heart. With a depth of more than 3.5 kilometers (2.2 miles), this Andes abyss is among the deepest chasms on Earth. As a point of comparison, this is more than twice the downward distance of the Grand Canyon.

Known locally as el cañón de Cotahuasi, the name is Quechua (the language of the Indigenous people) for "the Canyon of Wonders." This grand crevasse was formed by the Cotahuasi River as the flowing water etched its way through the valley between two peaks, Coropuna and Sulimana. Each apex towers over the ravine at the inspiring height of more than 6,100 meters (20,000 feet) above sea level. These awesome pinnacles are found about 200 kilometers (124 miles) north of the city of Arequipa.

These peaks were created by the upthrust of Earth's crust along the northwest end of South America's lofty Altiplano (high plane) around 10

million years ago. Geological forces began a wild bash at the confluence between the two mountain massifs and three rivers. Through much of the Tortonian stage, this corner of the planet was moshing like it was at a hard rock festival. Just as tympanic drums can highlight a musical passage, volcanic eruptions regularly echoed above the constant beat of erosion. Like the kick drum rumbling out the bottom sounds, there were continuing tectonic uplifts banging out weaknesses between the subducting Nazca Plate and the South American Plate. The drumroll of activity caused the land to fold and fault, creating a long chain of mountains—a cordillera—between the massifs. Glaciers chiseled into the din, dancing over the landscape, like fingers on a keyboard. This composition resulted in one of the deepest chasms in the world.

Wondrously, Cañón de Cotahuasi displays twenty-four stratigraphic units; metamorphic, sedimentary, and volcanic-sedimentary rocks are recognized, ranging from the Precambrian to the recent Quaternary. The oldest rocks in the canyon are approximately 1.5-billion-year-old, continent-supporting metamorphic rocks of the Basal Complex. The deeply incised canyons of the Andes demonstrate the net effects of environmental processes on the planet's shell in response to surface elevation. Of great scientific value, this example of topographic evolution through geologic time offers information on how tectonic forces, climate change, chemical and physical variations, and erosion impact a landscape.

Ammonite collectors will appreciate the Toro and Charcana districts of the canyon, which are each about an hour's drive from Cotahuasi. The marine fossils are found on the hillsides by Toro in the limestones of the Arcurquina Formation. Look in the upper part of the preservation, which is composed of craggy blue-gray limestone with a weathered yellow, red, and pink crust. The 100-million-year-old matrix is spotted with

Forget the Grand Canyon—Peru's Cotahuasi Canyon laughs at its "little brother" from a staggering 11,004 feet down. This monster gorge, sliced through the Andes by the relentless Cotahuasi River, sits between two mountain giants: Coropuna and Solimana. It's nature's way of showing off its carving skills.

Image credit: NASA Earth Observatory images by Joshua Stevens, using Landsat data from the U.S. Geological Survey and topographic data from the Shuttle Radar Topography Mission (SRTM). Story by Adam Voiland. Public domain, via Wikimedia Commons.

*Arnioceras* Hyatt, 1867
Order: Ammonitida. Appearance and lifespan: Early Jurassic, 196.5 to 189.6 MYA. Location: Atacama Desert, Chile. Size: 4 cm.

various coffee-colored chert nodules that are rich in preserved urchins and ammonites. This fossiliferous, thick gray limestone layer is well exposed in the valley area facing Tomepampa and Taurisma, as well as in the Pampamarca and Cotahuasi sectors. Here, you can find ammonites from several geologic periods.

The fossiliferous limestones hold cephalopods such as *Manuaniceras*, an Early Cretaceous ammonite that lived about 109 to 105 million years ago. It was a relatively involute subgenus of *Oxytropidoceras*; it had ribs that folded at the venter, creating enough space for a sharp keel. The endemic *Glottoceras raimondii* survived during the Early Cretaceous about 113 to 101 million years ago. A member of the ammonite superfamily Engonoceratoidea, this involute ammonite had pie-slice ribbing along its flanks, with ornamental crimping edging a narrow, depressed venter. Found in various locations around the planet, *Parahoplites* thrived during the Early Cretaceous, about 125 to 101 million years ago. A member of the ammonite superfamily Deshayesitoidea, this moderately evolute specimen had consistent ribbing that bifurcated mid-flank and joined across the rounded venter. Also look for the involute *Neolobites*, which can be found along with the banded heteromorph Anisoceratidae and the strongly ribbed *Calycoceras*.

*Santafecites santafecinus* d'Orbigny 1842
Order: Ammonitida. Appearance and lifespan: Early Cretaceous, 140.2 to 136.4 MYA. Location: Copiapó Province, Atacama, Chile. Size: 4 cm.

## Cotahuasi Subbasin Landscape Reserve

Similar in size to Grand Canyon National Park, this Andean scenic preserve contains twelve different landscapes. Ecosystems vary from temperate rivers at the bottom of the chasm to the cold arid desert climate found near the summits. The diverse biotas include many endemic medicinal plants with exotic names such as muña, retama, jara, kiwicha, and yareta. Staples like grains, corn, beans, and herbs are grown on agricultural terraces established before the Incan civilization dominated the canyon and built their empire.

The people of the Wari culture were the original stewards of Cotahuasi Canyon, but it was overrun by the expanding Incan realm in the thirteenth century. They mined the salt from the local repository at Huarhua and transported it along the legendary route of salt to Cusco, the capital of the Incan Empire. A half-hour drive in Cotahuasi Valley, Huarhua, is a picturesque pre-Columbian-era settlement that sits on a small volcanic terrace. There is a level of ten tombs, indicating that the salt mine was activated sometime in the sixth century. This historic hamlet is a folkloric pueblo with narrow cobblestone streets traveled by lamas and donkeys, whitewashed houses, and rustic balconies. Conveniently located, it is a short walk to the salt mine.

The people in the region still preserve their ancestral traditions. Communities of artisans ply trades like stone carving and carpet weaving. European influence can be seen in the colonial churches with locally influenced façades. Be sure to enjoy the majestic Sipia Waterfall, which flows off a cliff, cascading an awe-inspiring 150 meters (492 feet) before splashing on the rocks below. This stunning cascade is around 10 kilometers (6 miles) from the town of Cotahuasi. The easiest option is the regional *combi* van, an excellent and inexpensive way to travel around the canyon, that allows you to interact with the locals en route. It is also accessible by foot along paths less traveled; although the villages are relatively close to each other, the hike can be exhausting if you aren't acclimatized to these heights.

The easiest time of year to visit Cotahuasi is between May and November, when the days are sunny, the nights cool, and the rains scarce. Remember to bring bug spray.

## Leave from the Historic City of Arequipa

A journey to the canyon town of Cotahuasi is not for the faint of heart or the physically challenged. The journey begins at the Altiplano metropolis of Arequipa, the second most populated city in Peru. While you're here, appreciate why this metropolis is special. Founded in 1540, shortly after the Spanish conquistador Francisco Pizarro arrived in South America, the center of the city is a UNESCO World Heritage Site. Known as the Historical Centre of the City of Arequipa, it is a community of colonial-era statuesque buildings made of sillar, the whitish-pink volcanic rock that is common in the region. Unique stately homes, churches, and public buildings represent an integration of European and native building techniques and characteristics. A joint effort, designed by colonial masters and built by indigenous masons and Spanish-heritage criollos, the combination of influences is represented by robustly constructed walls, archways, and vaults

*Sonninia Mirabilis* Douvillé, 1879
Order: Ammonitida. Appearance and lifespan: Jurassic, 170.3, to 168.3 MYA. Location: Neuquen Basin, Patagonia, Argentina. Size: 5 cm.

with intricate baroque ornamentation on the façades. This singular community is sprinkled with courtyards and open spaces. La Ruta del Sillar, or the Sillar Route, takes you to the stone quarry on the outskirts of the city.

Several bus companies depart from the local terminal late in the afternoon to make the drive of approximately 400 kilometers. The journey takes around twelve hours, beginning on the Pan-American Highway and changing to unpaved roads after Chuquibamba, some 3,000 meters (10,000 feet) above sea level. Along this epic journey, you will see farmers herding packs of llamas or sheep. Your transport will climb to the hamlet of Vizca and traverse the sky-kissing pass between the peaks of Coropuna and Solimana. Once you have championed this crossing, there is the roller-coaster descent down, down, down into the wilds of the valley. It's a very impressive trip, day or night.

***Prionocycloides*** **Spath, 1925** (sp) split.
Order: Ammonitida. Appearance and lifespan: Late Cretaceous, 99.7 to 94.3 MYA.
Location: Aldama, near Chihuahua City, Mexico. Size: 9 cm.

## The Legalities of Fossil Prospecting in Peru

You can dig for fossils, go on excavations, and buy preservations in Peru, but you're not supposed to take them out of the country. The South American Republic is a treasure map full of astounding archaeological and paleontological sites, many of which are not well protected. Looters—known as *huaqueros*—are known to pillage these locales and export their plunders to foreign markets. The country's National Culture Institute has acknowledged that international demand has made fossil trafficking a growing problem. "Peru is a country rich in fossil beds, and we are seeing an increase in the trafficking of this wealth," confirmed Javier Vazquez, director of the institute.

At Jorge Chávez International Airport in Lima, customs agents have impounded thousands of pieces of vertebrate material such as megalodon teeth, penguin bones, whale incisors, mastodon jaws, and the like. Customs agents seizing the items have noted that they are "illegally obtained fossils" and considered part of Peru's "national patrimony." They will take your specimens, but there is no other penalty attached to the crime.

Do they want your invertebrates rocks? Probably not, as long as there aren't too many of them. Have some kind of documentation, if possible. Or roll them up with some clothing and mail them to yourself. The worst-case scenario is that the package doesn't arrive.

***Microderoceras birchi* Sowerby 1820** adult and juvenile.
Order: Ammonitida. Appearance and lifespan: Early Jurassic, 199 to 191 MYA. Location: *Birchi* Subzone,
*Turneri* Zone, Lower Lias, Charmouth, Lyme Regis, Dorset, England, United Kingdom. Size: largest is 10 cm.

# Size Matters between Ammonites

**A**mmonites were a quintessential model for sexual dimorphism. Did someone say sex? The term "sexual dimorphism" isn't nearly as suggestive as it may sound. Within the animal kingdom, sexual dimorphism merely means that males are a different size than females. Macroconchs represent the grander female forms, while the shrewdly labeled microconch refer to the smaller and often more intricately decorated males. Sexual dimorphism is virtually universal in cephalopods, both living and fossilized. "Macroconchs are the larger of the species, and they are female," confirmed cephalopod paleontologist Neal Larson. "This was firmly proven in a specimen of Hoploscaphites recently prepared. The specimen has an egg case with egg impressions in it."

## Lappets Were Erotic

Among *Homo sapiens*, the female traditionally developed features to better attract an appropriate male for reproduction. This is quite the opposite of the animal world. In nature, from birds to lions, males are typically the more vibrant and adorned creatures so that they may entice a female into mating. Among ammonoids, the mighty microconchs were often ornamented with sexy bells and whistles like protruding nodes and etched ribs, as if the bold ornamentation signaled the overall genetic quality of the male. Another evolutionary attempt to lure female ammonites were projections from either side of the male body chamber aperture like tusks. Known as lappets, they have been

Most commercially available ammonites with lappets are found in France.

**Neocosmoceras Blanchet, 1924**
Order: Ammonitida. Appearance and lifespan: Early Cretaceous, 145 to 139.8 MYA. Location: La Cadière, France. Size: 8 cm.

preserved among some Western European genera since the Middle Jurassic. One hypothesis is that this modification may have protected the animal. Another school of thought holds that lappets were a sign of maturity and fitness for breeding, particularly because these features are not present on the larger female ammonites. "Growing features such as big combat-ready horns required a lot of energy, and female animals already had to expend a lot of effort to produce eggs," observed Larson.

Lappets confirm two ammonite traits:

1. They had good eyesight.
2. They spent time in the euphotic zone.

Sexual dimorphism is apparent in this pair of male and female scaphites.

**Hoploscaphites spedeni (Landman and Waage, 1993)**
Order: Ammonitida. Appearance and lifespan: Late Cretaceous, 84.9 to 66.043 MYA. Location: Fox Hills Formation, Late Cretaceous, South Dakota, United States. Size: female is 12 cm, male 7.5 cm.

Image credit: Neal L. Larson

If we assume that these carapace swords had a purpose, they would not have been visible in the dark depths of the ocean. Lappets imply that ammonites lived in the sunlit layers of the seas, in shallower depths that were affected by daylight. Look for these projections on the evolute *Normannites*, a genus with hearty ridges on the flank of each whorl that yielded adult shells with a maximum diameter as large as a billiard ball.

## Fraternal Twins

Remarkably, male and female ammonoids weren't born with a size differential. Embryonic ammonitella eggs of both genders shared an identical origination and early growth experience. "From a developmental point of view, sexual dimorphism in ammonites can be described as two classes of individuals with juvenile development identical up to a size, from which two different morphotypes develop: a macroconch (female) and a microconch (male)," summarized Argentinian paleontology professor Horacio Parent, in conjunction with his Polish counterpart, Michał Zatoń.

The divergence came once the hatchling reached a critical diameter, and the maturing male and female ammonites diverged. Each developed the characteristic morphology and sculpture of their particular gender. Circling back to *Asphinctites tenuiplicatus*, juvenile growth was identical until the ammonitella diameter grew beyond 15 millimeters (.6 inch). Similar to puberty in human teenagers, sexual differentiation took hold and led to dimorphic development. As they matured, the robust microconchs remained tightly coiled, the aperture opening developed adornments with stronger ornamentation and more notable rib lines, and their whorls compressed. Conversely, macroconchs stayed conspicuously stout, and they developed dense contours, flattened sides, and a wide central depression.

Based on modern cephalopod life spans, experts estimate that most ammonoids likely lived one to five years. Smaller species probably reached maturity, spawned, and died within a year; larger ones most likely took more time. "All ammonites grew fast," confirmed Larson. "Cephalopods are among the most efficient animals in the world for converting food to growth." Among mature adults, there appears to be a higher percentage of macroconchs, implying that some males may have had the ability to impregnate multiple macroconch egg clusters.

**Ammonite Bite:** The complementary macroconch/microconch pairs are collectively known as antidimorphs.

The dotted area near the center of the photo is believed to represent an egg-carrying *Hoploscaphites* fossil.

***Hoploscaphites spedeni* (Landman and Waage, 1993)**
Order: Ammonitida. Appearance and lifespan: Late Cretaceous, 84.9 to 66.043 MYA. Location: Fox Hills Formation, Late Cretaceous, South Dakota, United States. Size: 11 cm.

Image credit: Neal L. Larson

## The Egg and the Coil

The tiny but mighty superorder *Plectronocerida* proliferated to become the faunal form from which all subsequent shelled cephalopods evolved. A creature with a conical, slightly arched conch, it appeared more than 500 million years ago. It set the precedent of secreting its shell within the egg capsule. "In the oldest ammonoids, the embryonic shells were considerably bigger and coiled less tightly than in later forms," noted University of Zürich paleontologist Kenneth De Baets.

*Mimosphinctidae* was an exemplary evolutionary early ammonoid for a variety of reasons. The

*Bactritoidea* was the first cephalopod subclass with the ability to generate large numbers of small eggs. Each of the roe held a round embryonic protoconch which developed several functional chambers prior to hatching.

### *Lobobactrites* (Schindewolf, 1932)
Order: Bactritida. Appearance and lifespan: Middle Devonian, 393.3 to 387.7 MYA. Location: Widder Formation, Southwest Ontario, Canada. Size: 3 cm.

adult was loosely wound with an open center. To accommodate its form, the female produced elongated protoconchs in an oval ovum. It spawned in small broods of two to three roe at a time. Shell shape extrapolated into a host of issues for the loosely looped juvenile ammonoids struggling to navigate a massive ocean. A failed experiment, *Mimosphinctidae* had ultimately disappeared by the end of the Paleozoic Era.

A more successful step up the evolutionary scale was the more sophisticated cephalopod subclass, *Bactritoidea*. This nearly straight-shelled cephalopod looked like a refined orthocerid, with a tapered conch. Bactroids are noteworthy for their ability to produce large numbers of small eggs, each protecting a round embryonic protoconch. As each ovum became more globular, it coincided with a decrease in embryo circumference. This gave the fertile female space to amplify the volume of roe that could be birthed.

Once the protoconch rolled, the new, snugly curled shell style changed the orientation of the larvae within the ammonitella to an upward-facing aperture. (In straight-shelled cephalopods, the opening faced downward.) Nature's attempts had led to the globular protoconch, which enabled cephalopods to spawn a mass of roe. Tightly spiraled conchs with a closed core proved to be a structurally sound design for survival. It allowed for more petite spawn, ultimately creating space for more eggs to develop. It was a win all around for the *Ammonoidea* subclass. "Although shape is not a proof of function, it appears plausible that juvenile conchs were selected rather for compactness while adult conchs were selected for improved hydrodynamic properties," noted renowned teuthologists Christian Klug, Kenneth De Baets and Dieter Korn. "This appears plausible because, at small conch diameters, swimming movements will not suffice for effective translocation, and a planktonic mode of life is likely."

For shelled cephalopods, it became the best of all oceans. The biomodification that yielded smaller embryos was coupled with a notable increase in adult size. The enlarged female conch had more body chamber volume, which resulted in even more prolific ammonoid reproductive activities. Dominant Late Devonian families like *Agoniatitidae*, *Mimagoniatitidae*, and *Gephuroceratidae* could birth an impressive 3,400 to 220,000 eggs with diameters as thin as a credit card. "The large number of offspring could have been the key to the rapid proliferation of the ammonoids in the aftermath of each mass extinction," offered De Baets.

Like salmon, ammonoids were typically semelparous animals—their grand finale was to lay or fertilize clutches with vast numbers of eggs. The reproductive process was likely followed by death. This breeding style allowed for the proliferation of young and strong cephalopods throughout the world's seas. The clutches of roe were fertilized by males, who also likely died after a spawning or a few. "Semelparity or iteroparity has previously been hypothesised for a number of fossil cephalopods, but usually this was inferred from a combination of life history traits," noted a study by the distinguished research team of De Baets, Klug, Landman, and Korn. "Taking all the evidence together, we think that a semelparous behaviour provides the best explanation but requires further supporting evidence." The researchers further noted, "The evolutionary relationships among early Palaeozoic cephalopods are still poorly understood and the published phylogenetic hypotheses quite different."

Egg masses may have been spawned on the sea bottom, amid plant life, and/or set afloat in mid-water. Once fertilized, the deposited ammonitella developed a tiny baby protoconch that consisted of an initial gas-filled chamber and a subsequent whorl. Upon hatching, the young offspring floated up toward the surface of the ocean, intermingling with the plankton, which they could feast upon at random. Research paleontologists have noted a growth discontinuity that corresponds to the ammonitella hatching out of its egg casing; they named it the nepionic constriction.

Although ammonoid larvae were likely too tiny to effectively maneuver through ocean currents, their shell functioned as a buoyancy apparatus as soon as they were born. They consumed plankton and grew rapidly, allowing the conch nacre to thicken naturally by the constant sway of the sea.

*Frechites occidentalis* (Smith, 1914) ammonoid group.
Order: Ceratitida. Appearance and lifespan: Triassic, 247.2 to 242.0 MYA. Location: American Canyon, Nevada, United States. Size: largest is 4 cm.

Endemic to northwestern Germany, *Parapuzosia seppenradensis* is the largest known species of ammonite. This is a cast of the original specimen.

**Parapuzosia seppenradensis (Landois, 1895)**
Order: Ammonitida. Appearance and lifespan: Late Cretaceous, 83.5 to 70.6 MYA. Location: Lüdinghausen-Seppenrade, Münster, North Rhine-Westphalia, Germany. Size: 1.8 m (5.9 ft) in diameter, although the living chamber is incomplete. It is estimated that the entire carapace would have been 200 cm (6.6 ft) in diameter.

Image credit: Markus Schweiss, CC BY-SA 3.0, via Wikimedia Commons.

## Go Big

The size and shape of the shells in utero was one of two evolutionary trends that facilitated the coiling of embryonic *Ammonoidea*. The other was that some species of hatchlings were able to grow into massive adults. Assess ammonoid fossils from the Devonian, Triassic, and Early Jurassic periods, and it becomes apparent that the shell rarely reached a size exceeding a dinner plate. Then the miracle of evolution took hold among some families, and ammonites grew to nightmarishly large sizes: think giant squids. From the Late Jurassic through the Late Cretaceous, the seas were occupied by incredibly large and disconcerting species such as *Parapuzosia seppenradensis.* Unearthed mostly in the fossilized lagoon environs around

Westphalia, Germany, examples of this somewhat involute shell with slightly rounded sides regularly spanned more than 2 meters (6.6 feet) or more. Its relative, *Parapuzosia bradyi*, the largest documented North American example, was almost as large.

## We Visit Ammonites: The Haute Provence Geopark

**Name: Haute Provence Geopark**
**Location: France**
**Age: Carboniferous to Late Eocene**
**Early Jurassic Ammonites**
**Date of UNESCO Inscription: 2004, 2015**

*Dalle* is the French word for "slab," and it's what the locals call this remarkable location near the town of Digne-les-Bains. Perched on the side of a hill in the Alpes-de-Haute-Provence prefecture of southeastern France are a series of limestone clay beds. There is one particular gray slab that slopes at a 60° angle and has an irregular surface that looks like it has stuff stuck to it. Upon close inspection, it is composed of a rather homogeneous aggregation of more than 1,550 ammonites. Officially known as the Les Isnards Ammonite Slab of Digne-les-Bains, it is situated in the Alpen foothills at an altitude of around 650 meters (711 feet). This extraordinary block of exposed cephalopods already covers an area the size of a spacious suburban home, and there is more to be excavated. The exposed massive mass mortality wedge—some 320 square meters (3,444 square feet)—captures the effects of climate change. Ninety percent of the ammonites in the Dalle are a mass mortality of *Coroniceras multicostatum*. A member of the family *Arietitidae*, members of this aesthetic group are easy to recognize. They are evolute with prominent rib marks and a keel separated by a pair of grooves that run along the

Lake Sainte Croix in the Alpes-de-Haute-Provence, France, is actually a half-century-old reservoir fed by the Verdon River.

Image credit: Marc DUPUY, CC BY-SA 2.0, via Wikimedia Commons.

venter. Bigger and more sophisticated than its predecessors, the genus has complex ammonitic sutures as well as an individual anaptychus with a striated surface. *Coroniceras* could go large—it was the ammonite that dominated seas around the globe in the Early Jurassic.

Typically, there is more species diversity in cephalopod-laden formations yet the only other genus found sparsely interspersed is *Angulaticeras*, a member of the family *Schlotheimiidae.* It was one of the more prominent ammonite types prior to the rise of *Coroniceras.* Researchers have noted that the ammonite shells and accompanying bivalve and echinoderm remains do not indicate any preferred orientation: so, if there had been marine currents, they had no impact.

Remarkably, some 198 million years ago, this hillside was the seafloor of a relatively deep basin. Sediments indicate that two separate geomorphological events forced this slab up and into its distinctive lithospheric position. These happenings also created the European Alps and were the catalyst that closed ocean access to many shallow continental basins.

*Coroniceras* **Hyatt, 1867**
Order: Ammonitida. Appearance and lifespan: Early Jurassic, 182 to 175.6 MYA. Location: Millau, Aveyron, Occitanie, France. Size: largest is 36 cm.

The ammonite wall at Digne-les-Bains in the Alpes-de-Haute-Provence region of France features almost exclusively the species *Coroniceras multicostatum*.

Image credit: Amélie Pataud, CC BY-SA 4.0, via Wikimedia Commons.

The slope of this assemblage of ammonites is one of the beguiling aspects of the ammonite wall at Digne-les-Bains, France. Another is the rare occurrence that this herd of Ammonoidea is composed primarily one species.

Image credit: Pmau, CC BY-SA 4.0, via Wikimedia Commons.

The Dalle is perhaps the most distinctive site of the massive Haute Provence Geopark, one of the original natural wonder locations chosen by UNESCO. Protecting the region's Alps in southern France, the open space includes sixty municipalities, uniting the district's environmental, geological, anthropological, and philosophical heritage. Capturing a 300-million-year journey through our planet's history, the Alpes de Haute Provence area serves as prototype for United Nations branded ecotourism.

## Digne-les-Bains

Active since neolithic times, Digne-les-Bains now positions itself as a town devoted to ammonites. Inspired by their world-renowned *Coroniceras Dalle*, the town feels like a cephalopod celebration. Artisans honor the ammonite by replicating its gracious spiral in bread, chocolate, ceramics, etchings, and photographs, and escargot is a specialty at local restaurants.

**Ammonite Bite:** The Europeans offer elaborate recipes for cephalopod bread, but here are simple, American-style directions for baking ammonite rolls:

1. Buy a tube of breadstick dough.
2. Open the package and unroll the precut prepared lengths.
3. Twist the stretch of pastry to create some texture.
4. Roll the length of dough into a spiral, so that it resembles an ammonite shell.
5. Bake according to the directions on the package.

Voilà! Enjoy.

Today, Digne-les-Bains is a hamlet boasting 20,000 residents. It is the capital of the Alpes-de-Haute-Provence region, nestled in the Bléone Valley, about 140 kilometers from Marseille and

*Cardioceras cordatum* (Sowerby 1813)
Order: Ammonitida. Appearance and lifespan: Late Jurassic, 161.2 to 155.7 MYA. Location: Villers-sur-Mer, Calvados, France. Size: 7 cm.

152 kilometers from Nice. The township is strategically located at the convergence of three rivers: the Mardaric, the Eaux-Chaudes ("Hot Waters"), and the Bléone. It is a vacation village of chapels, churches, museums, chateaus, magnificent views, and charming boutique hotels. More than half the population is comfortable enough not to work. The rest of the residents engage in tourism or agricultural activities, particularly sheep raising.

## We Dig Ammonites: Aveyron, France

Even though fossils may not be extracted from the Geopark, there are other opportunities. So much of the country was once underwater that France is a wonderland for *Ammonoidea* collectors. Locals say you can find pyritized cephalopods throughout *la Patrie*. Uncovering relics may be as simple as exploring a tractor trail on a dead-end road in the middle of nowhere. Poke around in the right plot of dirt and, within minutes, you may find a handful of fossils. A favorite gathering experience is near

Aveyron, a medieval French city about four hours west from Digne-les-Bains and two hours northeast of Toulouse.

The ammonites of Aveyron are about 10 to 20 million years later than those in the Geopark, and there are a variety of genera to be found. Here, you may discover the rimmed and ribbed *Hildoceras*, the spiny *Porpoceras*, or a highly desirable *Lytoceras*—which may have Saturn-like rings spaced around its coils. They can be easy to spot if you are in the right place at the right time. Search after a good rain, and you may discover Ammonoidea that have weathered out of the hillside and are lying on the ground. For more pristine specimens, pick at the lithified marl.

Beyond ammonoids, Aveyron is chockfull of architectural fascinations like churches, castles, and monasteries. Local produce markets take place on Saturdays and offer regional treats like Roquefort cheese.

Near the small city of Millau is the place to find *Coroniceras*, the genus on the Digne ammonite slab. It is also the home of the tallest bridge in the world, which has a structural height of 336.4 meters (1,104 feet). For anthropology fans, 125 kilometers (78 miles) to the west is Martiel, an area with many megalithic, dolmen tabletop tombs.

## We Attend a Fossil Show: Sainte-Marie-aux-Mines

A medieval village known for its rich cache of minerals, Sainte-Marie-aux-Mines is a fossil acquisition experience like no other. Nestled in eastern France in the Val d'Argent ("Valley of Silver"), its mountainsides are laced with tunnels leading to concealed treasures. Each year around the summer solstice, the quiet hamlet of Sainte Marie hosts Euro Mineral, one of the largest and most impressive fossil markets in Europe. During the four-day affair (the first two days are for professionals only, so get your credentials in order in advance), there are conferences and workshops, and a selection of exotic fossils and minerals with a regional flare. Hundreds of vendors drive their rocks to this ancient community, and it transforms into a bazaar of paleo mineralogic prizes, the likes of which you've never seen before. Tents line the medieval cobblestone lanes, which aren't much wider than a horse cart. Sales tables are set up in venues wherever dealers can find space: in the gyms and lunchrooms of esteemed centuries-old schools like Sainte Geneviève and the Lycée Polyvalent. Sellers are found in the crooks and courtyards

of venerable timeworn churches like Sainte Madeleine. Have an idea of what you're looking to buy. Ammonoids and other conservations are on view throughout town, and you may feel euphorically overwhelmed by the variety, location, and irony of the farmer's market ambience.

Sainte-Marie-aux-Mines is part of the resource-rich upper Rhine region, in proximity to Germany and Switzerland. Mining archaeologists surveying the terrain have counted more than one thousand different quarry entrances, galleries, shafts, and well openings dating as far back as the tenth century. Once a major ore-yielding hydrothermal seam, the town's many veins are exhausted. Fortunately, a few historic tunnels are still visible in the rugged landscape. An inventory of minerals was recently conducted, and the results were quite interesting. In the Hautes Vosges highlands, they recorded more than fifty primary metallic mineral deposits formed in the Paleozoic, with just as many aggregates formed during the Mesozoic. Most curious however, are the more than thirty different newly formed minerals that developed in vintage mining chambers after operations had ceased. The region's many reserves include familiar names like aragonite, silver, arsenic, chalcopyrite, galena, and sphalerite, along with exotic aggregates with names like tennantite, tyrolite, and weilite.

## We Dig Triassic Ammonites: Quiche Anyone?

The Lorraine region of France is about two hours northwest of Sainte-Marie-aux-Mines, and it is the absolute epitome of French; think vichyssoise and quiche, *s'il vous plait*. Strategically located at the crossroads of four nations, the area has a history of regal intrigue and marriage, political treaties, and religious zealots, and pâté. Culturally and tactically speaking, the land has played a transformational role in European history. Geologically, it is the

These two Early Jurassic **Hildoceras** ammonites are similar, yet different, in both species and preservation. The specimen on the left has been replaced by limonite with hematite, while the ammonite on the right has pyrite replacement.

(*left*) **Hildoceras bifrons (Bruguiere, 1792)**
Order: Ammonitida. Appearance and lifespan: Early Jurassic, 182.7 to 174 MYA. Location: Larzac Region, France. Size: 5 cm.

(*right*) **Hildoceras Hyatt, 1867**
Order: Ammonitida. Appearance and lifespan: Early Jurassic, 182.7 to 175.6 MYA. Location: Aveyron, France. Size: 5 cm.

French location of the very incredible Muschel-kalk—a segmented lithostratigraphic unit that covers large swarths of western, central, and northern Europe. It is a delight for ammonite collectors. The name is German for "shell-bearing limestone." In French, it's called *calcaire coquillier*. This central series of ceratite-laden sedimentary rock strata formed some 240 million years ago. Today, outcrops of conch- and carapace-dense limestone, marl, and dolomite are found along the Rhine in locations like Soultz-sous-Forêts, 40 kilometers (25 miles) northeast of Strasbourg.

The ceratites of the Lorraine Muschelkalk are a variation on their deep-ocean counterparts of the same stage. There is clearly a common ancestor, but ammonite species radiated like a mutating virus; they adapted and proliferated to fill environmental voids. Perhaps the best-known example from this fine digging area is *Ceratites nodosus*. Partial to open, shallow, subtidal marine environments, this hearty-shelled creature had prominent ribbing and grew as large as 30 centimeters (12 inches). *Ceratites semipartitus* was a sophisticated ammonite that matured to a size 25 percent larger than its counterpart. Hydrodynamically designed, it was large and discus-shaped with a narrow keel, and its shell diameter grew to more than 40 centimeters (16 inches). These ammonites lurked around reefs just beyond the tideline. They were often found with the fist-sized *Ceratites evolutus*,

Here is a classic example of direct development: the juvenile resembles the adult.

***Schloenbachia varians* (Sowerby, 1817)** (sp)
Order: Ammonitida. Appearance and lifespan: Late Cretaceous, 99.6 to 93.5 MYA. Location: Kolbay section, Mangystau Region, Kazakhstan. Size: largest is 7 cm.

## The Legalities of Fossil Prospecting in France

Article 552 of the French Civil Code states that everything above and below the ground belongs to the owner of the land. Wherever you dig, get some kind of documentation to show customs that you had permission to be where you were collecting fossils for personal use. Without this kind of proof, taking fossils can be considered the theft of movable property.

**Ammonite Bite:** When searching on private lands, do request the owner's permission before trespassing or digging.

named after its morphology of highly visible coils. *Ceratites enodis* looked like the spiral you would draw. It is small, smooth, simple, and very sexy— and the species list goes on for another dozen or so. What is notable about the *calcaire coquillier* specimens is that they are preserved as internal molds, so the lobed suture pattern unique to ceratites should be visible.

Throughout the region of Lorraine are several horizons laden with these history-marking cephalopods. The *Ceratites pulcher* zone is clearly defined, as are the under and overlying boundaries. The other strata are not as noticeably demarcated. Less common ammonoid catches may include *Ceratites munsteri, Ptychites studeri*, or *Balatonites balatonicus*. This grouping was adaptable, and examples are randomly strewn throughout formations, from deep-sea environments to marginal marine deltas, lagoons, and tidal flats. Also abundant in the *calcaire coquillier* are various brachiopods and oysters. Snake vertebrae and other bone bits may also be found, but they are rather rare.

Zero in on your locations via an online search. As you are roaming around, however, keep in mind that an easy place to find unburied treasures is around Europe's construction projects and quarry areas. Explore industrial sites, housing developments, or roadworks, and take a moment to appreciate the stratigraphy.

Ammonite with scree in its living chamber.

*Hauericeras* Grossouvre, 1894 (sp)
Order: Ammonitida. Appearance and lifespan: Late Cretaceous, 70.6 to 66 MYA. Location: Outhriate section, Ma'an Governorate, Jordan. Size: 27 cm.

# What Ammonites Ate

**L**ooks may be deceiving. Ammonoids were preserved as magnificent fossils, but when these marine mainstays were alive, they were no angelfish. Research reveals that these cephalopods were by and large aggressive predators. Essentially, *Ammonoidea* preyed upon any sea creature that was much smaller than itself. Analysis of stomach contents from several families reveals a smorgasbord of seafood. An ammonite diet might include any and all combinations of minute marine life: zooplankton, crinoids, starfish, shrimp, other molluscs, fish, crustaceans, as well as whatever other tiny aquatic animals might be available. Like all cephalopods, ammonoids were opportunistic feeders and were even believed to have cannibalized hatchlings and other cephalopods.

Planispiral Ammonoidea typically preyed on the cornucopia of microfauna in the epipelagic zone. They were voracious hunters with the ability to conquer most depths. It is easy to envision ammonoids as they hovered around reefs, lingered over seabeds, or lurked just below the surface in search of plankton and other prey.

Phragmocone detail of a Late Cretaceous ammonite with various faunal elements (including ammonitella and fish bones) trapped inside the living chamber of its shell. Such captured-in-time moments make any observer wonder if this detritus was part of an ingestion process, or had it merely stagnated within the ammonite conch?

*Hauericeras* **Grossouvre, 1894** (sp) detail.
Order: Ammonitida. Appearance and lifespan: Late Cretaceous, 70.6 to 66 MYA. Location: Outhriate section, Ma'an Governorate, Jordan. Size: 27 cm.

Ammonoids were known for eating anything smaller than themselves.

*Metalegoceras sundaicum* **(Haniel, 1915)**
Order: Goniatitida
*Neoproetus indicus* **Tesch, 1923**
Order: Proetida. Appearance and lifespan: Permian, 272.5 to 265.0 MYA. Location: Maubisse Formation, West Timor, Indonesia. Size: goniatite, 25.6 cm.

> **Ammonite Bite:** The ability to be omnivorous undoubtedly contributed to the ammonite's nearly 400-million-year life span.

## Dining Parameters

The small diet theory compliments the ammonoid jaw shape. With its shield-like aptychus and a larger lower jaw, *Ammonoidea* morphologies were not well designed for catching and cutting up large prey. If the target organism was too hefty to suck through its system, this crafty cephalopod used its ring of ten arms to seize the unsuspecting creature. Poetically, this masterful scuba hunter manipulated its limbs, directing the meal into its awaiting maw—the buccal mass. This collection of mouthparts transformed the prey from creature into a pulpy wad. After being jawed by the aptychus, the meal was then shredded by the radula, a raspy feeding structure that was feathered with comb-like teeth conveniently nested along the ammonoid's feeding tract. Brutalized, beaten, and tattered into something consumable by the creature's buccal mass, the meal shifted to the digestive system of the prehistoric cephalopod. If energy was not immediately needed, the sustenance progressed into the crop for storage; otherwise, it was transferred to the stomach to complete ingestion.

Perchance this scaphite died of from bivalve gluttony?

**Discoscaphites conradi (Morton, 1834)**
Order: Ammonitida. Appearance and lifespan: Late Cretaceous, 70.6 to 66.0 MYA. Location: *Hoploscaphites nicolletii* Zone, Trail City Member, Fox Hills Formation, Dewey County, South Dakota, United States. Size: 5 cm.

*Lithacoceras* ammonite with aptychus and poop (also known as *Lumbricaria*).

**Lithacoceras Hyatt, 1900** (sp)
**Lumbricaria Münster, 1831**
Order: Ammonitida. Appearance and lifespan: Late Jurassic, 151 to 145 MYA. Location: Solnhofen Limestone, Germany. Size: 4.5 cm.

## Poop Has Proof

"Coprolite" is the scientific name for fossilized poop. More specifically, *Lumbricaria* is the trace fossil name for ammonite *coprolite*. Scientists have found more than a few little squirts in the cephalopod-rich Solnhofen Limestones of Germany, the Jurassic Coast of England, and other ammonoid-abundant locations. What does a stringy little pile of preserved poop tell us about the creature that excreted it? Let's start with the diameter. The Solnhofen extrusion is not very wide, only about the thickness of a credit card. This leads one to figure that the organism evacuating it was not particularly large. If it was found in a pile or in curved filaments, we can assume that the creature had time to expel its sustenance leisurely as opposed to dealing with the constant concern of becoming prey.

Technical analysis indicates that ammonite coprolite was predominantly calcite (try saying *ammonite coprolite* five times fast). Paleontologists have also found bits of a free-floating crinoid named *Saccocoma* in its dung and digestive tract. These plant-like animals may have been an ammonite comfort food. One of the more aggressive genera of the Late Jurassic, *Aspidoceras* had a more robust and powerful radula than its other ammonoid relatives, yet it too was found to have snacked on crinoids.

Close-up of aptychus and coprolite. Pet trainers profess that a dog will not poop where it eats, but perhaps this is not true of ammonites.

*Lumbricaria* Münster, 1831
Appearance and lifespan: Late Jurassic, 151 to 145 MYA.
Location: Solnhofen Limestone, Germany. Size: aptychus, 2 cm.

## Starved Out

Picture a plethora of ammonite hatchlings happily floating around in the euphotic zone some 66 million years ago. Now imagine the Chicxulub asteroid randomly ramming into the planet—*KERPOW!*

**Ammonite Bite:** Because plankton and other microorganisms were staples in the ammonoid diet, the demise of ammonoids may have freed up the ocean for modern plankton feeders like the blue whale, one of the largest mammals known to have existed.

The fragile young buoyant ammonites that were peacefully riding the waves were decimated, along with their food source. The tragedy that beset these cephalopods affected much of life on Earth. In the proverbial blink of the geologic eye, three-quarters of all species of life on our planet were eliminated. From giant sauropods to ammonite babies, it all disappeared. Little taller than a crocodile survived. "Food resources may have been the dominant cause of their extinction," accounted Royal Mapes, paleontology professor at Ohio University and research associate at the American Museum of Natural History.

## Ammonite Holiday: The Iberian Peninsula

A virtual history of cephalopod shells may be found dispersed throughout southwestern Europe. Should you search, you will find that Spain and Portugal have in situ dig sites that span more than 350 million years of marine history. The Iberian Peninsula was once lowlands bathed by shallow seas, and it offers the collector the luxury of choosing the age of ammonites they wish to gather. It offers tremendous fossil-viewing and excavation situations.

The catalyst for this period-busting accumulation was the mighty supercontinent Pangea. Time and tectonics rifted the massive sheet of land. As the Triassic progressed into the Jurassic,

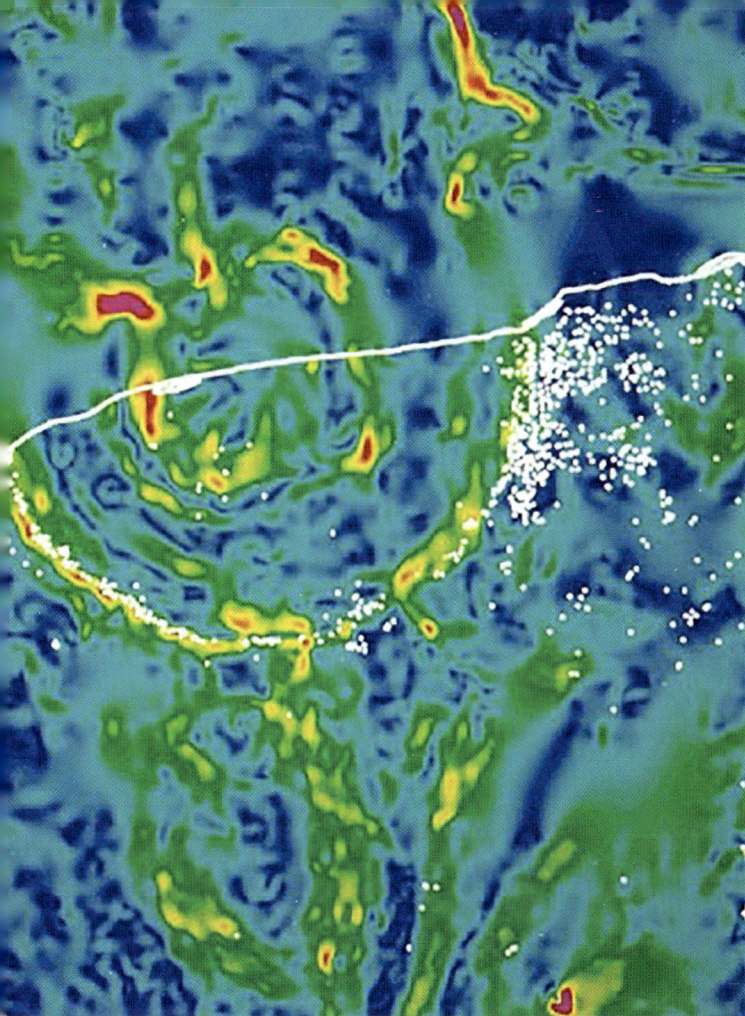

This prominent series of concentric rings, which straddles the Gulf of Mexico at the Yucatan Peninsula, pinpoints the contact zone of the Chicxulub asteroid impact.

Image credit: Public domain, via Wikimedia Commons.

*Phymatoceras anomalum* (Merla, 1932)
Order: Ammonitida. Appearance and lifespan: Jurassic, 182 to 175 MYA. Location: Torcian Limestone, Monte Cucco, Perugia, Italy. Size: 6 cm.

the northern continent of Laurasia tore away from its motherland. Consequently, new arm of the Tethys Ocean flooded in to fill the gap between landmasses. But that was just the start. Next, the western slice of Pangea tore in a north-south direction, opening a rivulet that would grow into the Atlantic Ocean. Ammonite species became the biostratigraphic markers that demonstrate these passages were linked earlier than previously thought.

## We Visit Ammonites: The Southwest Alentejo and Vicentine Coast Natural Park

**Name: Southwest Alentejo and Vicentine Coast Natural Park**
**Location: Portugal**
**Age: Upper Devonian to Cretaceous**
**Date of UNESCO Submission: January 31, 2017**

The marine connection between the Atlantic and Tethys oceans is preserved around the southwestern tip of continental Europe at Ponta do Telheiro.

The angular unconformity at Praia do Telheiro confirms an aquatic connection between the Atlantic and Tethys oceans.

Image credit: André Cortesão, CC BY 3.0, via Wikimedia Commons.

This corner of Portugal captures and confirms the breakup of Pangea and the development of the Atlantic Ocean some 183 million years ago. "The western-central regions of the Iberian Peninsula were temporally or permanently flooded by the Protoatlantic," explained Manual Segura, Fernando Barroso-Barcenilla, Pedro M. Callapez, and Javier Gil, members of a research group from Universidad de Alcalá de Henares in Spain. "The main depositional episodes and faunal assemblages were related to global sea-level changes . . . influenced by local tectonics and continental influx episodes."

At Ponta do Telheiro, there are conservations that verify that ammonites from the cooler Boreal Sea migrated south into the Mediterranean Basin at a time when the bodies of water were supposedly separated. An ideal example of a cephalopod that made this journey was the hearty *Lytoceras fimbriatum.* You'd know this genus if you saw it.

The ammonite had a shell that curled very much like a tight ram's horn. It was often etched with striations, and the exterior coil was sometimes encircled with Saturn-like rings. They could have diameters beyond the width of a basketball. This family was particularly prolific in Europe, where Lytoceratidae ammonites flourished from the Early Jurassic through the Cretaceous.

This southwestern corner of the European continent is a Jurassic depot for the study of paleontology. The approximately 195-million-year-old biota of the Algarve Basin is the only place you will find *Ptycharietites*—another type that has a ram's horn whorl. It was strengthened by sawtooth lobe and saddle sutures that pay homage to its ceratite ancestors. *Reynesocoeloceras elmii* was a rare member of the *Dactylioceratidae* family that was whorled like a serpent. Delightfully spiraled with extensive skinny ribbing, it proliferated alongside another exclusive species, the thin-walled *Juraphyllites helveticus.* From the order Phylloceratida, the exterior of its smooth involute shell usually exhibits fine growth lines, but its sutures were complex—branched with rounded tips.

At the north end of the park, around the volcanic rocks of Vila Nova de Milfontes, is an excellent area to visit older ammonoids in situ in Iberia. Dating from approximately 362 to 307 million years ago, this Paleozoic unit reveals a vast fossiliferous content with goniatites, brachiopods, trilobites, and crinoids. Embedded in the matrix are involute ammonoids like *Arnsbergites, Lusitanoceras, Paraglyphioceras,* and *Dombarites.* Ask the locals how to find them hidden in the rocks, waiting to be spotted like four leaf clovers in a field.

These cephalopod locations reside in the countryside along the Atlantic Coast at the bottom of Portugal, an area that is easy to appreciate. The awe-inspiring Southwest Alentejo and Vicentine

(top) **Dombarites liratus Ruzhentsev and Bogoslovskaya, 1971**
Order: Goniatitida. Appearance and lifespan: Early
Carboniferous, 326.4 to 318.1 MYA. Location: Kiya River,
Orenburg Oblast, Russia. Size: 3 cm.

The example of **Lytoceras** (*left*) shows the shell stripped of its
outer layer, revealing the suture patterns. On the bottom is the
specimen with its external nacre.

(left) **Lytoceras Suess, 1865**
Order: Ammonitida. Appearance and lifespan: Early Cretaceous,
113 to 100.5 MYA. Location: Betioky, Atsimo-Andrefana,
Madagascar. Size: 16 cm.

Coast Natural Park spans 150 kilometers (93 miles)
of rugged cliffs and sandy coves that have captured
approximately 362 million years of Earth's history.
Beyond the geological and paleontological delights
of the fossiliferous cliffs, modern-day mammals
flourish in these protected environs, watering at
flora-lined ponds. At the coast, the sea teams with
schools of sharks—blues, makos, the occasional
hammerhead—as well as fevers of stingrays.

## We Dig Iberia

About a half-hour inland from the nature park, near
the picturesque Rota Vicentina, lies the provocative
cephalopod-laden Bordalete Formation. The site is a
thick aggregate of dark shales interlayered with thin
layers of siltstone dotted with ammonoid nodules.
Because the Carboniferous ammonoid *Muensteroceras* is mixed with ancient arthropods, scientists
have dated this formation at approximately 350

**Eocanites Librovich, 1957**
Order: Prolecanitida. Appearance and lifespan: Carboniferous, 360 to 353 MYA. Location: Kugutyk Formation, Orenburg, Russia. Size: 5 cm.

radiated forty-three different genera that produced about 1,250 species, supplying the rootstock traced to all later Mesozoic ammonoids. With some time and dedication, it's not too difficult to find fossil nodules that have weathered out of the cliff face.

Further up the coast is a spot that locals say is one of the most pleasant places to collect Jurassic fossils in Portugal. The Lourihnã Formation is about an hour north of Lisbon. Known for its Early Jurassic sauropod footprints and vertebrae finds, it's also a wonderful place to discover ammonites. There's the special and endemic *Epophioceroides*, evolute and ribbed with insets on either side of its keel. *Ptycharietites* was smoother with a narrow venter. The widespread genus in this formation is the well-known, handsomely ribbed *Asteroceras*.

## Look But Don't Dig Ammonites: Andalucía

Iberian ammonites rarely come to market so collect them if and when you are able. While you're on the south end of the peninsula, head east to take advantage of the location and check out the outcrops and pleasures of Andalucía, Spain. Positioned at the extreme south of Europe, this region

million years old. Hidden inside some of the Carboniferous concretions are swirled surprises like *Becanites* and *Eocanites*. Both were members of the relatively small and stable order Prolecanitida. These ammonoids had a lineage that spanned more than 100 million years. Although not as diverse as their goniatitid contemporaries, Prolecanatida

### The Legalities of Fossil Prospecting in Portugal

When determining what to do with fossils you have discovered in Portugal, be aware that the country has "a complete absence of laws" when it comes to preservations. According to collectors and conservationists, recent efforts to protect Portugal's extensive fossil preservations have "fizzled out."

Local rockhounds suggest having paperwork from area museums. They also advocate proper prospecting behavior. "Anyone coming to Portugal is to be the best steward they can be of objects that are best understood in their original context. Thus finding something as float or as part of an earthwork is different from finding something that's still encased in an outcrop. So, not to belabor the point, people should be responsible and use their best judgement . . . report anything of interest" to the local museum.

is a fertile Eden flanked by the Atlantic Ocean and the Mediterranean Sea. Beyond its strategic location, the autonomous community's fruitful land contains rich deposits of minerals that have served civilization since prehistoric times. This paradise is like no place else on the planet. An oasis that has served as one of the crossroads of the world for centuries, Andalucía is an amalgam of cultures. Part Arabic—with exotic Moorish arcades—and measurably Catholic—a province of grand Renaissance churches, the area is a mix of Europe, Africa, and the Middle East that has established itself as a unique culture within southern Spain. The location is home to seven World Heritage Sites, the people are kind, and the food is simply delicious. The lifestyle has its own rhythm. Days are hot and humdrum, but the region comes alive at night. The sounds of castanets and the tango twang of acoustic guitars flow into the streets. Lovers and tourists grow euphoric from the enticing scent of orange blossoms, especially after they've sampled the delights of the tapas bars.

As you grasp the lay of the land, you'll notice that the cities of Antequera and Málaga are separated by a small mountain range known as the Sierra del Torcal. Hidden away in these hills is a Jurassic wonderland sprinkled with ammonites, Mesozoic delights, and anthropological secrets. Around the time of the Cretaceous–Paleogene extinction, an upthrust of limestone rock lifted this spectacular cephalopod-laden formation from the briny deep, like the proverbial phoenix rising from the ashes. It elevated these benthic beds to a height of more than 1,300 meters (4,265 feet) above sea level. After seemingly reaching the sky, the limestone was carved and cleaved by erosion, freezing, and thawing. The result is a forest of flat-topped karst hoodoos—a 150-million-year-old, cephalopod-dappled relic from the marine corridor between the present-day Atlantic Ocean and the Mediterranean Sea.

The Tornillo formation at El Torcal.

Image credit: Edmundo Sáez, CC BY-SA 3.0 ES , via Wikimedia Commons.

**Ammonite Bite:** Karst acts as a large sponge, storing rainwater and releasing it within the rock, which encourages the limestone to dissolve. This process erodes into sinkholes, chasm-etching streams, caves, springs, and other characteristic features.

## The Route of the Ammonites

### El Torcal

**Antequera Dolmens Site Ensemble**
**Location: Antequera, Spain**
**Age: Jurassic**
**Date of UNESCO Inscription: 2016**

Like the names in Spain, locations may be long and descriptive—take the Walk of the Ammonites at the Antequera Dolmens Site Ensemble in the Sierra del Torcal Mountain Range, as an example.

An ammonite embedded in a cliffside along the Route of the Ammonites in El Torcal de Antequera UNESCO site.

Image credit: Alexmofer99, CC BY-SA 4.0, via Wikimedia Commons.

Instead of its formal title, Europeans save their syllables and affectionately refer to it as El Torcal, and enjoying the ammonites in the region is a wonderful adventure. This inspiring site is located in the province of Málaga, off the A45 road in Andalucía. Beyond a bevy of ammonites, this vibrant setting has two natural monuments and three Bronze Age cultural sites. If you are fascinated by the planet and/or its inhabitants, you'll no doubt come away more enriched by your experiences in southern Spain.

All healthy cephalopod enthusiasts will enjoy the Monday ammonite hike. This three-hour guided tour of the formation and its fossils follows a rugged 4.5-kilometer (2.8-mile) loop trail. Be sure to make a reservation through the Torcal Alto Visitor's Center. During your walk, you will scrutinize the hillsides of marine fossils. Look for *Nebrodites* (also known as *Passendorferia*). Known as the Mediterranean Perisphinctidae, it is distinguished by a wide center with visibly squarish whorls adorned with ribs that forked on the flanks. This Jurassic genus was usurped by the family Ataxioceratidae, which prevailed along the northern epicontinental margins.

If you can pull your microfocus away from the ammonites and accompanying belemnites and look up and out, it's possible to appreciate the impressive panoramic views. There are the hoodoos of the El Torcal Park, and the equally impressive landscapes of southern Spain and the African coastline.

There are right seasons and wrong seasons for enjoying the Walk of Ammonites. Spring or autumn are the winning times of year——as it is neither too hot nor too cold. The elevation dictates warm waterproof clothing. There also may be fog and/or strong north winds. Sturdy trekking shoes are strongly recommended due to the irregular and steep terrain. Bring water.

For additional viewing pleasure around El Torcal, there are a plethora of flowering plants, including lilies, wild rose trees, red peonies, and thirty varieties of orchids. Overhead you may find

**Ammonite Bite:** This area of Spain has minimal light pollution, allowing for extremely favorable conditions for stargazing. The El Torcal Astronomical Observatory often hosts sky-watching evenings for exploring the universe.

*Nebrodites (Passendorferia) birmensdorfensis*
**Brochwicz-Lewinski, 1973**
Order: Ammonitida. Appearance and lifespan: Late Jurassic,
161 to 155 MYA. Location: Cherves, France. Size: 8 cm.

majestic falcons, vultures, and eagles, likely hunting large reptiles. On the ground, you might see the Iberian ibex and perhaps a family of bulls, among other critters.

While in the neighborhood, you can stimulate your anthropological and geological whimsy at the Antequera Dolmens suite of monuments. A dolmen is megalithic tomb with a large flat stone lintel laid across standing rocks. They are found only in select locations around Western Europe. Dating as far back as the Neolithic period, the Menga Dolmen, the Viera Dolmen, and the Tholos of El Romeral are pre-Christian catacombs camouflaged as mounds in a rugged landscape. These obscured stone tributes built by early humans remained undiscovered for thousands of years before becoming landmarked anthropology locations.

## The Legalities of Fossil Prospecting in Spain

When in Spain, find out about prospecting at the local office of the ministry of industry, trade and tourism, and be prepared to take no for an answer. Spanish Historical Heritage Law No. 16–1985 states that cultural property is subject to export control. "Movable or immovable property of a historical nature that can be studied using archaeological methodology forms part of the Spanish Historical Heritage, whether or not it has been extracted and whether it is to be found on the surface or underground, in territorial seas or on the continental shelf. Geological and paleontological elements relating to the history of man and his origins and background also form part of this heritage."

Breach of conditions is subject to confiscation of fossils and penalties and fines.

That is the official line. Local collectors reveal that invertebrates found on the Spanish mainland are not typically a problem if you want to take them out of the country. Spain is one of the twenty-seven European Schengen Area countries that have officially abolished passports and many types of border control. That leeway does not apply to Spain's Balearic Islands, where rocks are likely to be confiscated at the airport security check.

The ornate Middle Jurassic *Stephanoceras* ammonite has a name meaning "crown horn" in Greek. A marine chambered palace of calcium and time, this specimen would be even more grand if the living chamber were complete.

**Stephanoceras Waagen, 1869**
Order: Ammonitida. Appearance and lifespan: Middle Jurassic, 170.9 to 168.2 MYA. Location: Fresney, France. Size: 12 cm.

The extensive Alpine Orogenic System was created by convergent movements between the African Plate, the Arabian Plate, and the Indian Plate from the south;, the Eurasian Plate from the north;, and many smaller plates and microplates. Activity along this intercontinental seam began in the Early Cretaceous and continues to this day.

Image credit: Jo Weber, public domain, via Wikimedia Commons.

## We Dig Ammonites: The Baetic Cordillera

During the Late Mesozoic Era, continental fragments shifted positions, and the Tethys Ocean slowly disappeared. The changes forced the African and Iberian plates to collide, elevating marine deposits to terrestrial heights. The Baetic Cordillera is but one of an extensive line of ranges that plate tectonics have pushed up along the southern margin of Eurasia. Recognized in part by the extensive array of aquatic fossils that surfaced in its wake, the Alpine-Himalayan orogenic belt stretches for more than 15,000 kilometers (9,300 miles). This ribbon of crustal squeezing has continued for tens of millions of years and is still actively building rock sequences. If you're a rockhound, that means there are always new fossil alcoves to be found.

**Ammonite Bite:** After the Pacific Ring of Fire, the Alpine-Himalayan orogenic belt is the second most seismically active region in the world, generating 17 percent of the world's largest earthquakes.

The Baetic Cordillera segment of this escarpment begins in western Andalusia and spreads over much of the bottom corner of the Iberian Peninsula, stretching to the isles of Mallorca and the Rock of Gibraltar. The geologic shifts have built lands holding an extraordinary variety of Jurassic and Cretaceous cephalopods. Between Granada and Jaén are a couple of dozen locations where you

*Crioceratites loryi* **Sarkar, 1955**
Order: Ammonitida. Appearance and lifespan: Early Cretaceous,
132.9 to 129.4 MYA. Location: Drome, France. Size: 5 cm.

can find an assortment of preserved marine life of different ages.

Different pockets hold different treasures. In a Cretaceous layer lies desirable genera such as the loosely coiled *Crioceratites* or *Acantholytoceras*; both are members of the heteromorph superfamily Ancyloceratoidea. The surface was banded with fine ribs, and if you're lucky, you'll find one with spines along the venter. Nearby may be *Berriasella*, another member of the Perisphinctidae family. This disc-shaped evolute ammonite likely had a compressed whorl section and ribs that extended across a narrowly rounded outer rim. Not too far away, peering out of a Jurassic outcrop could be a large *Emileia*, which had a finely ribbed body chamber that resembled a flattened barrel. Also to be found are several members of the Stephanoceratoidea superfamily, whose name is Greek for "crown horn." These evolute ammonites may be spoked, and noded with a ribbed keel reminiscent of a motorcycle tire.

As a related aside, the next time you're in Madrid, check out the subway station at Ciudad Universitaria. The black limestone that lines the platform walls is dotted with ammonites.

*Homoeoplanulites* **Buckman, 1922** with lappet.
Order: Ammonitida. Appearance and lifespan: Middle Jurassic,
167.7 to 164.7 MYA. Location: Bedoulian, France. Size: 7 cm.

**Ammonite Bite:** The species *Berriasella jacobi* is the index ammonite used to define the base of the Cretaceous Period, 145.5 million years ago.

Ammonoid with bite marks highlighted by drusy quartz and calcite.

***Metalegoceras* Schindewolf, 1931**
Order: Goniatitida. Appearance and lifespan: Permian, 290.1 to 272.3 MYA. Location: Laktutus, Cova Lima District, East Timor. Size: 4 cm.

# What Ate Ammonites

A t the end of the Paleozoic Era, ammonoids were high on the food chain, masters of their domain at the peak of their reign. True to life at the top, there was nowhere to go but down. Mesozoic evolution had *Ammonoidea* usurped by larger and more advanced predators. Their position on the trophic level became one similar to today's squid—simultaneously serving as both feared predator and potential tasty morsel. Researchers have hypothesized that, by the dawn of the Jurassic, a primary reason that ammonoids produced so many young was merely to serve as escargot for larger, more complex carnivores.

## From Predator to Prey

What were these evolved beasts that dined on ammonoids as if they were clams on the half shell? The answer depends on the location. Just as shelled cephalopods consumed a variety of creatures, a bevy of bigger animals slurped the soft bodies of these curling molluscs as if they were enjoying raw oysters. The Jurassic biota of Europe's Solnhofen Archipelago was teeming with marine vertebrates that made Ammonoidea part of their diet. The bony fish order *Pycnodontiformes* had teeth capable of breaking ammonoid conchs, but even more interesting are the theories that these creatures were cannibalized by other members of their cephalopod brethren—the squid-like

Goniatite with predation holes at the back of the living chamber.

**Discoclymenia cucullata (von Buch, 1832)**
Order: Goniatitida. Appearance and lifespan: Late Devonian, 364.7 to 360.7 MYA. Location: Jebel Ouaoufilal East, Errachidia Province, Drâa-Tafilalet Region, Morocco. Size: 3 cm.

A stealth hunter fed on this ceratite.

**Ophiceras Griesbach, 1880** (sp)
Order: Ceratitida. Appearance and lifespan: Early Triassic, 252.17 to 247.2 MYA. Location: Lidak, East Nusa, Tenggara, Indonesia. Size: 7 cm.

coleoids. Numerous Mesozoic ammonoid conchs have been preserved with bites uniquely located at the posterior of the body chamber. A chomp to this area disengaged the muscles, allowing the predator to draw the carcass, probably still alive, out of the test.

Examine ammonites preserved in the Muschelkalk in the area around Nusplingen, west of Munich, Germany. You'll discover that around three-quarters of the Perisphinctidae and Oppeliidae family members exhibit these unique bite marks. Around this region that was once part of the Tethys Sea, the sole family of calamari that performed this operation was the ghoulishly named Vampyroteuthidae, along with Trachyteuthididae. Both families had species with an appropriately large body size, capable arms, and

jaws strong enough to tackle some ammonites. Straight-shelled belemnites like *Hibolithes* would have been a likely addition to the killing crew. Conversely, the stocky tests from the local *Aspidoceratidae* fossils are regularly found with their shells crunched—likely by less delicate hunters in search of a meal.

Along England's Jurassic Coast, it was beaked squids that preyed on ammonites. Studies by the Yorkshire Geological Society have established that nearly 1/5 of the spiraled cephalopod fossils found have bite mark–like punctures near the back of the largest chamber. Ammonite preservations from the English Channel are often found with this section broken off, likely the result of an aggressive death attack.

Ammonite with predation bite emphasized by drusy quartz cluster.

***Asteroceras* Hyatt, 1867**
Order: Ammonitida. Appearance and lifespan: Early Jurassic, 196.5 to 189.6 MYA. Location: Black Ven, Charmouth, Dorset, England, United Kingdom. Size: 8 cm.

## Jumping the Shark

Randomly discovered in a European collection was an *Orthaspidoceras* ammonite pierced by a tooth from the Early Cretaceous shark, *Planohybodus.* The slender and pointed tooth was the sort usually associated with fish flesh eaters. The anatomy of these long-extinct cartilaginous predators didn't really make this type of ammonoid an ideal choice of prey. Emerging in the Jurassic, the cephalopod *Orthaspidoceras* had a thick, involute shell that could have grown as large as a plate. A further deterrent was that their few whorls were likely adorned with a row of gnarly nodes, which were not easy on the mouth.

Were ammonite-slurping sharks a Mesozoic reality? *Planohybodus* was new to the seas, so perhaps it was a learning process. If the cephalopod pierced by the cartilaginous fish tooth is not an anomaly, it demonstrated a sly strategy on how to slaughter these seemingly delectable, coiled molluscs. Any perforation of the shell would have had an impact on an ammonite's buoyancy control. Even a small hole devastated the siphuncle system; the puncture would have allowed water to flow uncontrolled into the conch, ultimately impeding maneuverability. Once the shark eliminated the possibility of the ammonite's escape, all it had to do was use its jagged, spikey teeth to pluck the meat from the shell, and voilà, it is *Calamars à la mer.* Paleontologists have found regurgitated Mesozoic shark meals with ammonoid conch bits in the mix. It makes evolutionary sense; many modern cartilaginous fishes have sturdy teeth that can crunch the hard exteriors of shelled animals.

## Mosasaurs

Mesozoic oceans may have had different creatures, but they were less exotic than you might expect. If one were to describe an animal as having a "powerful, hydrodynamic build, strong flippers, and teeth meant to ensnare fish," you could be describing either a great white shark, a killer whale, or a mosasaur.

In the Cretaceous, while dinosaurs roamed forested floodplains dotted with rivers and lakes, saline seas played host to a significant group of marine reptiles—the *Mosasaurus.* Looking vaguely like a monitor lizard, they were long, sleek, dark, scaly-skinned, air-breathing saurians that likely evolved from reptiles that slithered into the waters to feed on marine life. Around 100 million years ago, what may have once been lizard legs evolved into the mosasaur's muscular flippers. They had a

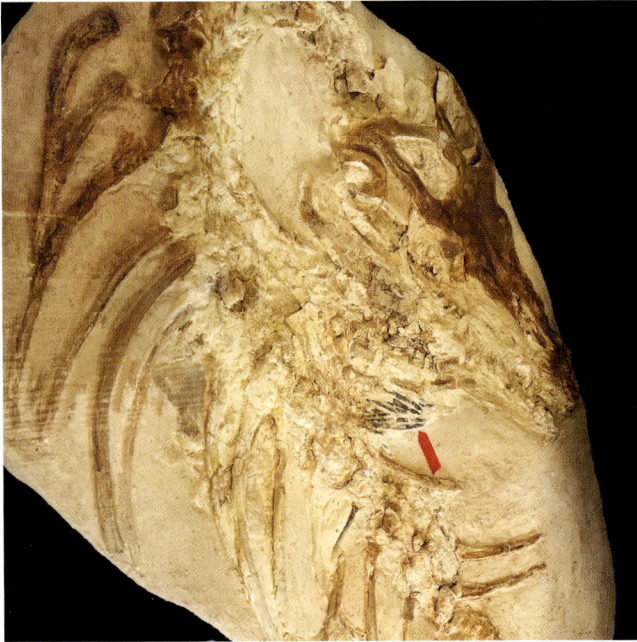

Mosasaur fossil with belemnites in its stomach, on display at the Keystone Gallery, Scott City, Kansas, United States.

Image credit: Courtesy of Neal L. Larson, edited

This *Placenticeras* ammonite displays precise bite marks generated by the jaws of the mosasaur genus *Plioplatecarpus*. Various genera of these aquatic reptiles consumed ammonites, and some, such as the genus *Globidens*, featured mushroom-shaped teeth that had evolved expressly to puncture cephalopod shells.

**Placenticeras costatum (Herrick & Johnson, 1900)**
Order: Ammonitida. Appearance and lifespan: Late Cretaceous, 83.5 to 70.6 MYA. Location: Bearpaw Shale, *Baculites reesidei* Zone, Korite Quarry, Saint Mary River, Lethbridge, Alberta, Canada. Size: 36 cm.

Mosasaur jaw **Globidens Gilmore, 1912**
Order: Squamata. Appearance and lifespan: Late Cretaceous, 89.8 to 83.6 MYA. Location: Kinney County, Texas, United States. Size: 28 cm.

*Placenticeras* image credit: Courtesy of Neal L. Larson, Schaffert Collection, edited.

dragon-strong paddle tail capped with a rudder-like fluke that easily propelled them through the seas. Their double row of teeth and powerful, loosely hinged jaws were perfectly designed to hunt ammonites, along with almost anything else they might crave. The pterygoid teeth were smaller and located adjacent to the primary tooth, toward the center rear of their mouths. The larger set of chompers were primary teeth designed and customized for grasping prey. Genera of large Cretaceous ammonites such as *Pachydiscus* and *Placenticeras* have been found with triangular patterns of holes in their shells. The size of the punctures and their presence on both sides of the conchs corresponds to where teeth are located in upper and lower mosasaur jaws. The consumption process was to bite the ammonite shell gently enough to weaken it. This action likely disengaged the muscles that held the ammonite soft body to the conch.

Free from its protection, the animal was easy to slurp down, like a piece of sushi. This dining strategy allowed the mosasaur to extricate their prey while ingesting a minimum of shell fragments. It's believed this marine reptile's method was to attack

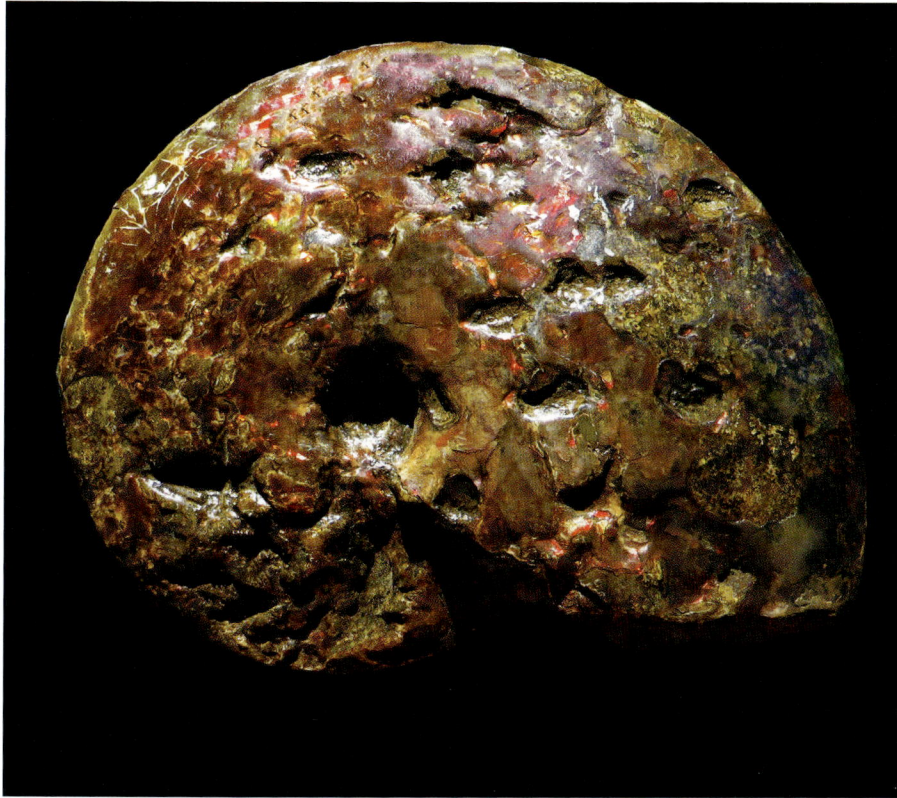

This *Placenticeras* ammonite has been dubbed the *War Horse* because of the array of mosasaur bites, which may indicate it was prey in a battle between multiple marine reptiles. The shell bears dramatic evidence of the fatal encounter.

*Placenticeras costatum* **(Hyatt, 1904)**
Order: Ammonitida. Appearance and lifespan: Late Cretaceous, 83.6 to 72.1 MYA. Location: Pierre Shale, Niobrara County, Wyoming, United States. Size: 30 cm.

its target from above, so death was as surprising as a thunderclap.

Mosasaurs were apex predators—sea monsters with a foreboding presence. They ate virtually everything in their path. Stomach content analysis of these reptiles has revealed bony fish, sea turtles, plesiosaurs, sea birds, and belemnites, often accompanied by a side order of ammonites. Some species had diets that were more restrictive to their environment.

## Alp Dirt

If you've spent time wandering the European Alps, you've probably noticed that some of the peaks skew at unique angles. Skiers and climbers find themselves aggressively leaning into the hillside to compensate for the deformation. It makes you wonder: how did they develop such a peculiar centrifugal twist? The answer is likely because the

Alps grew in several phases, through a series of traumatic tectonic events. Each bend holds a story from a different piece of geologic history.

The foundation for this European mountain chain began some 770 million years ago, when a protrusion developed in Earth's crust. Over the next 470 million years, it seeped a symphony of schist, gneiss, and limestone. This rock show eventually became the base of the Alps. Then, like a scherzo for the brass family in the third movement, metamorphic granite intruded into the western sector late in the process, reinforcing the foundation for rising ridges.

Evolution is a tricky process; it's utopian in vision without the incalculable variables, disasters, and extinctions that may be its side effects. The young Alps battled ongoing marine erosion and the constant aggregation of sedimentary rocks. The Alpine Tethys Ocean—which had been the lube between the African and Eurasian plates—eventually

***Douvilleiceras clementium* (d'Orbigny, 1841)**
Order: Ammonitida. Appearance and lifespan: Early Cretaceous, 112.03 to 109 MYA. Location: Courcelles, Aube, France. Size: 8 cm.

disappeared. Without a buffer, these two geographic masses scraped and scrapped over which would subduct the other. This tectonic confrontation led to creasing in Earth's mantle that built mountain peaks and canyons. The process continued for tens of millions of years. Along the way, the African landmass cracked, calving the Adriatic Plate during the Cretaceous Period. This tiny but mighty piece of geotectonics crawled over the other morphologies, bulldozing magma up against the Alps and carving significant thrust faults. This repeated activity, in conjunction with other topographic activities, contorted the slopes that would become world-renowned ski resorts like the Matterhorn, Zermatt, and Mont Blanc. Landmass movement and the subduction of the African Plate is at least partially responsible for volcanoes and earthquakes in southern Italy, where the city of Pompeii was covered in volcanic ash from the eruption of Mount Vesuvius in 79 CE. And did you know that the city of Napoli (Naples) is built on the caldera of a volcano? It's all related. Bully for paleontologists in that all this activity has built a European landmass that has entombed a diverse fauna of transitional aquatic and terrestrial reptiles that swam along and sometimes ate the ammonoids.

**Ammonite Bite:** Geophysicists believe that the Adriatic Plate is still moving independently of the Eurasian Plate in a north-by-northeast direction with a slight counterclockwise rotation. The fault zone running through the Alps that captures this game of lithospheric bumper cars is known as the Periadriatic Seam.

Enjoy the view of Lake Lugano from Morcote, Switzerland.

Image credit: Valser, public domain, via Wikimedia Commons.

## We Visit Tethys Sea Ammonoids: Monte San Giorgio

**Name: Monte San Giorgio**

**Location: Italy and Switzerland**

**Age: Middle Triassic**

**Roughly 243–239 million years ago**

**Date of UNESCO Inscription: Switzerland 2003,**
   **Italy 2010**

Always a poetic setting, whether woody and green or dappled in snow, Monte San Giorgio gently inclines out of Lake Lugano, straddling the border between Italy and Switzerland. There is much to enjoy around this pyramid-shaped mountain, including one of the finest fossil records of Middle Triassic marine life. An unsuspecting place for a treasure trove, Monte San Giorgio reaches a moderate 1,097 meters (3,599 feet) in height and is laden with fairy tale–perfect chestnut woods. There is a steep rise from the lake and a gentler slope toward the southern light. On the exterior, Monte San Giorgio looks remarkably similar to

the other Alps that surround the lakes of northern Italy and Switzerland, but this mount of the Lugano Prealps recorded a prolific ecologic unit. Preserved in time are five fossil levels of life and climate evolution roughly ranging from 243 to 239 million years ago.

"Organic-rich laminated shales and limestones from Monte San Giorgio are known as famous fossil Lagerstätten for excellently preserved fossils from the Middle Triassic Period. The various bituminous shales are thermally immature and rich in diverse organic compounds, which provide unique substrates for active soil microbial communities," explained Sania Arif of the Institute of Microbiology and Genetics at Georg-August-Universität Göttingen, Germany.

Once upon a time in the Early Mesozoic, the south end of the Alps was a shallow sea located at the western margin of the Tethys Ocean. The landscape was an archipelago of small islands and carbonate platforms, similar to the modern-day British Isles, the Bahamas, or the Hawai'ian Islands. The difference is that these European mounds were built by a prolific genus of calcareous algae called *Diplopora*. Like evil aliens, the dulce invaded, segregating the nearshore environment by forming an offshore reef that grew into a barrier against the open ocean. It created an isolated 100-meter-deep (328 feet) lagoon that cultivated the diverse and distinctive ecosystem of Monte San Giorgio. In this calm cove, a rich aquatic fauna developed. Beyond ammonoids, there was a unique biota of sea life that included brachiopods, molluscs, crustaceans, echinoderms, conodonts, and an array of fish. The alien-looking reptilians were versatile, with some species comfortable on both land and sea. The shoreline was thick with tree ferns, conifers, primitive plants, insects, and other little creatures. This snapshot of marine and terrestrial life became entrenched in an anoxic seabed of fine blackish mud, flushed with bacteria. Through

*Reineckeia* **Bayle, 1878**
Order: Ammonitida. Appearance and lifespan: Middle Jurassic, 161.2 to 145.5 MYA. Location: Haute-Marne, France. Size: 10.5 cm.

[approximately] 241.2 million years ago," noted Rudolf Stockar, the geosciences curator of the Cantonal Museum of Natural History in Lugano, Switzerland. "Most of the spectacular vertebrate fossils [reptiles and fishes] together with important index fossils, including ammonoids and daonellid bivalves, come from this formation."

Ceratite horizons include layers named for *Eoprotrachyceras curionii* and *Protrachyceras archelaus.* Both are members of the superfamily Trachyceratoidea. This group fashioned involute shells, often highlighted by a furrowed keel, that largely covered the preceding whorls. An interesting and desirable collectible, they were known to reach diameters of more than 10 centimeters (4 inches), with their large flank often ornamented with ribs and nodes. Another layer is named for the ceratite *Arpadites arpadis*, which is appreciated for its flat, round, evolute shell adorned with central nodes and radial ribs. It may be found in the Meride Limestone, as well as at many Italian fossil-collecting sites, along with a similar-looking ceratite genus, *Kellnerites* and the evolute *Nevadites*. In all, more than one hundred species of *Ammonoidea* and other invertebrates have been recorded, including Triassic insects. Paleontologists have also identified around twenty-five species of reptiles; fifty types of fish; and many genera of flora, particularly conifers. In all, scientists from Italy and Switzerland have excavated a cache of approximately 21,000 fossils and counting.

The incredible fossil accumulation of Monte San Giorgio was revealed when mines were dug to extract oil from bituminous shale deposits during the second half of the nineteenth century. Excavators were awed when their exploration uncovered thousands of quality preservations, including several large vertebrate species. "Specimens from the Monte San Giorgio area are considered to have inhabited reef environments," said Stockar.

time, the area was compressed by the weight of accumulating sediments, which facilitated the detailed fossilization of these oil shales. The result is a prolific paleontology resource that reveals the wonders of species evolution in the face of geoclimatic changes.

## Meride Limestone

In a formation that's as thick as the tallest skyscrapers in the world hides a time capsule of ancient sea life. For our interests, the most important fossil beds of Monte San Giorgio are the central sections of Triassic sedimentary marine carbonate limestone and dolomite, punctuated by ash pockets. "A volcanic ash layer lying a few meters below this boundary yielded a minimum age of

*Trachyceras* **Laube, 1869**
Order: Ceratitida. Appearance and lifespan: Late Triassic, 237 to 228 MYA. Location: Hosselkus Formation, Lake Shasta, California, United States. Size: 10 cm.

*Properisphinctes Spath,* **1931** geode.
Order: Ammonitida. Appearance and lifespan: Jurassic 161 to 156 MYA. Location: near Liesberg, Jura, Switzerland. Size: 4.5 cm.

"*Saurichthys* (a ray-finned fish) is regarded as an ambush predator hunting in the water column. Among the ichthyofauna of the studied section, it is unquestionably the top predator." In the mix were plenty of placodonts—lizard-like, barrel-bodied marine reptiles with top-flat teeth custom-designed for feeding on shelled invertebrates, preferably squid and belemnites. Ammonoids were not high on their dining list, according to Swiss and Italian paleontologists.

Area scientists have been studying the fossils found on Monte San Giorgio since the location was exposed by the oil miners. The first journal on area fossils, *Pachypleurosaurus, Monte San Giorgio*, was published in Lombardy, Italy, in 1854 by naturalist Emilio Cornalia. He established a lifetime association with the preservation. As the director of the Museum of Natural History in Milan, he oversaw the first Italian scientific excavation at the Besano Formation location. After World War I, Switzerland became involved with the part of the site in their country. That led to the University of Zürich

conducting more than fifty open-air digs over the next half century.

This Triassic Alp is in the elbow of the two branches of Lago Lugano and is cradled by a canton of unique Italian Swiss–influenced villages that straddle the boarder of both countries. Whether you're in the slope-side village of Meride or Serpiano, or at the summit of Monte San Giorgio, there are magnificent views to be had. The Swiss side of the alp was inaugurated as a UNESCO World Heritage Site in 2003, with the Italian territory added in 2010. The United Nations chose the site because it "is the single best-known record of marine life in the Middle Triassic period, and records important remains of life on land as well."

An accessible paradise, the area is an easy 1.5-hour drive north of Milan. Perhaps stop for a cappuccino in the medieval city of Monza, an automotive mecca that's home to the Italian Grand Prix. The ultimate destination is Meride, a picturesque township appreciated by geologists, paleontologists, oenologists, adventurers, and polenta

Italian ammonoid cast with blackened suture pattern.

*Calliphylloceras nilssoni* (Hebert, 1866)
Order: Phylloceratida. Appearance and lifespan: Early Jurassic, 183 to 176 MYA. Location: Passo del Furlo, Macerata Province, Marche, Italy. Size: 8 cm.

connoisseurs. A throwback from another age, this medieval village has treasures hiding around almost every one of its many corners, in its courtyards, and on its hillocks.

To get to the peak of Monte San Giorgio, you can glide over the forest to lake level via the oldest automated cable car in Switzerland or hike one of several trails. Ask the concierge at your hotel for details, but be sure to get to the top because the scenery is spectacular. Relax on the sky terrace; enjoy a beverage; and savor the panoramic view, which spans from the Lepontine Alps to the Apennines.

The trek back down is exceptional, too. Follow the yellow signs indicating the route to Meride. As the trail weaves around to the Italian side of the monte, you'll find a geo-paleo path with illustrated panels. In Italophile style, your first landmark will be marked number 9, but here you will be in front of one of the initial mines excavated on the slopes.

Drilled in the early twentieth century, there are about five tunnels, each around about 400 meters (1,312 feet) in length. At this spot, bituminous schist was extracted and transported to Meride to make exotic medicinal oils and ointments. Your next notable paleo point is Acqua del Ghiffo. In 1931, Bernhard Peyer discovered a distinctive, large flippered reptile that became known as *Ceresiosaurus calcagnii.* His finds are now on display at the University of Zürich, where he was a professor of paleontology and later the dean. You'll know you're getting toward the end of your hike through paleo history when you see a factory chimney. Operated between 1910 and 1950, this is where the schist was transformed to meet the needs of the populus. But the fun isn't over just yet. Look for the panel for Val Mara—the spot of four different excavation sites for ammonites, fish, crustaceans, land plants, and the debut of two insect species. In another few minutes, you will reach Meride. Hike well done.

Switzerland has a patchwork quilt of languages and fossil exportation laws, and the legal ammonite sites are few and far between, but they do exist. There is an authorized site in the north, in the vicinity of Laufen (the headquarters of the company that makes Ricola herbal throat drops). The site is Jurassic and features ammonites such as *Malladaites* as well as sea urchin spines and brachiopods. Mine around online for location details.

## We Dig Ammonites: Capo San Vigilio

"Legally, Italy draws no distinction between the theft of artifacts, fossils, paintings, or grave robbing, all are held as cultural treasures . . . But having said that, don't be too worried to pick up some rocks when you're in Italy. Be very careful to avoid the off-limits localities (like Monte San Giorgio). And, if possible, ask permission when you're seeking on a private property," said Italian fossil

collector Niccolo (no last name please). "If you're looking for a way to fly the fossils you've collected out of Europe, leave from a country where you haven't been digging."

An ammonite search is a delightful reason to ramble around the Italian Lakes district. Meander through paradise to Lago di Garda, a leisurely 2.5-hour drive from Monte San Giorgio. As you approach, you will be awed. Formed by glaciers at the end of the last Ice Age, this is the largest lake in Italy. It's also a popular holiday destination, ideally located between Milan and Venice on the edge of the Dolomite Alps.

In this bucolic setting lies Capo San Vigilio, a notable ammonoid location because of a Middle Jurassic tectonic unconformity called the Aalenian succession. This conservation captures an interval of a rising marine transgression accompanied by an aboveground surge in humidity, which ultimately resulted in the aggregation of *Ammonoidea* and other shell remains in the shallow areas. The geologic memento of this dried body of water is a layer-cake stratigraphy displaying three distinct time frames.

The uppermost section is dubbed the San Vigilio Oolite. More than 65 million years younger than Monte San Giorgio, it has a characteristic lithology of pinkish yellow grainstone with mineralized granules—also known as ooids, thus oolite. San Vigilio is a magnificent snapshot of ammonite evolution from approximately 174.1 to 170.3 million years ago. The debut of the involute *Leioceras comptum* denotes the first of three designated ammonite horizons. It is topped by *Erycites fallifax*, its shell was spiraled in thick evolute whorls ornamented by ribbing and two rows of well-developed knobs along the edges. The top tier of this Aalenian triumvirate is highlighted by *Ancolioceras opalinoides*—an involute species with a razor keel and suture lines that looked like maple leaves.

This trio of ammonoid layers had an extensive supporting cast of cephalopods and other sea life, with each zone preserving its own unique biota.

Have you noticed, when Italian is translated into English, it often becomes a delightful phrase that an anglophone would not have used? Take the approximately 170-million-year-old stratum that sits atop the oolite layer. Maps name it the "Calcari a *Skirroceras* del Capitello," which English speakers would translate as the "*Skirroceras* Limestone Wayside Shrine." Area paleontologists refer to it as "Acqua del Chiffon." Once you get there, you'll discover members of the ammonite superfamily *Stephanoceratoidea*. These evolute Middle Jurassic creatures typically had a shell of coiled whorls lying on top of one another, augmented by a rounded keel. The conchs were ornamented with prominent ribbing, complex suture lines, and a textured aptychi. Prevalent sister taxa in the limestone include *Riccardiceras* and *Westermannites*.

*Phymatoceras elegans* **Merla, 1932**
Order: Ammonitida. Appearance and lifespan: Early Jurassic, 182 to 175.6 MYA. Location: Piobbico, Pesaro, Italy. Size: 5 cm.

## The Legalities of Fossil Prospecting in Switzerland

The rules of collecting fossils in Switzerland are perplexing. There is a dearth of official information about prospecting and keeping fossil cephalopods obtained in the Swiss Confederation. What is obvious is what to do with items from shell deposits on the beach. The country's branch of the World Wildlife Association noted, "Apart from corals and a small number of snails and shells (see Giant Clams), marine animals found on beaches do not fall within the scope of international provisions." Calcite, aragonite, shale, or limestone: is a shell a shell?

"The issue is that it depends on the canton," explained Christian Klug of the University of Zurich. "In Ticino, it is generally forbidden. In Solothurn, you can get registered and then search for fossils. In Aargau, when it's only invertebrates, it is not a problem. I don't know the rules of all cantons like you probably wouldn't know for each state in the U.S.

"Reason should apply: You shouldn't use heavy machinery and explosives without asking for permission. For quarries and private land, ask for permission. I would say it would be best if people contacted the local museums; they will know."

*Calliphylloceras* **Spath, 1927**
Order: Phylloceratida. Appearance and lifespan: Early Jurassic, 182.7 to 174.1 MYA. Location: Passo del Furlo, Macerata Province, Marche, Italy. Size: 6 cm.

The uppermost massif of the Aalenian succession has the lyrical name of the Rosso Ammonitico Veronese (Red Ammonites of Verona). It is an ode to sea life from approximately 168 million years ago that has been captured in red rock. The place to pick the rosy limestone facies of this succession is near the delightful Tuscan hamlet of Sassorosso. Located 2.5 hours south of Lago di Garda and 45 minutes west of Bologna, this charming village sits on an outcrop 993 meters (3,258 feet) above sea level and overlooks the surrounding valleys. This tiny settlement of barely 100 individuals is proud of their 22 buildings. Virtually the entire village was constructed with locally quarried modules of Rosso Ammonitico limestone. Look for fossils in the building blocks.

Head to the outcrop immediately downhill from the village, near the active quarry. Here, the sediments are characterized by rhythmic alternations of reddish nodular limestones and thin rosy marls and shales with easy-to-spot *Arietities* ammonite concretions surfacing from the matrix.

**Ammonite Bite:** Several Jurassic sedimentary bodies in the Southern Alps, Central Apennines, and Sicily are called Rosso Ammonitico. Some noble local paleo fans have added the region name to facilitate identification, so don't be surprised if you see areas with names like Rosso Ammonitico Lombardo and Rosso Ammonitico Veronese.

These *fossil* knobs come in a range of shapes and sizes. Some are like elongated ovals larger than a man's fist; others are almost as round and as small as a marble. At the Rosso Ammonitico massif, you will notice that the original shell of the ammonite fossil has been dissolved, and specimens are preserved via a calcite internal cast. The bonus to this preservation is that you get to enjoy the suture lines.

***Stephanoceras* Waagen, 1869**
Order: Ammonitida. Appearance and lifespan: Middle Jurassic, 171.6 to 168.4
MYA. Location: Kilchberg, Switzerland. Size: 10 cm.

**Ammonite Bite:** The homes of Sassorosso often have horseshoes hung over their doors, with the open part of the horseshoe hanging down. Local folklore has it that, as you enter under the lucky charm, you will be showered by good fortune and spread it to those around the house. As you leave, prosperity rains down upon you, bringing excellent luck on your travels.

While blemished ammonoids with mineral replacement dominate the macrofauna, there are other cephalopod treasures. Aptychi, belemnites, and brachiopods are in view along the quarry surfaces, waiting to be picked.

## The Legalities of Fossil Prospecting in Italy

Italy is another locale whose laws about fossil exporting are difficult to pin down. The *Bel Paese* has been labeled a "no collecting zone." Because of Italy's many anthropological treasures, there has been a blanket rule that valuable or mundane fossils uncovered in the Italian territory are the property of the Italian state. But according to Italian translators and attorneys, this interpretation seems to be in flux. An article added to Italy's Code of the Cultural and Landscape Heritage refers to "things that interest paleontology." This wording has initiated a conversation about things "that do not interest" the study of paleontology. But what is and what is not of importance has yet to be defined. An English translation of content supplied by the Italian Society of Malacology notes that generally speaking, "[t]he possession of fossil shells, recovered freely on the ground [and this effectively excludes unauthorized excavation and the consequent removal of specimens]and preserved for study, is tolerated by the authorities." For current information about the situation, reach out to the Civic Museum of Natural History in Milan.

The heteromorph ammonite *Oxybeloceras* has been aptly described as having a conch shaped like a paperclip.

**Oxybeloceras meekanum (Whitfield, 1877)**
Order: Ammonitida. Appearance and lifespan: Late Cretaceous, 83.5 to 70.6 MYA.
Location: *Didymoceras cheyennense* ammonoid zone, Pierre Shale, Boulder County,
Colorado, United States. Size: 6 cm.

# Heteromorphs Are a Wacky Part of Evolution

Hold one ammonoid or seven hundred; whether they're globular or more disc-shaped, there's a circular familiarity to them. These spiraled shelled sorts are known as homomorphs because members of the same species resemble each other. Each conch was constructed of a flat coil, and every whorl was adjacent to the curve before it, as prescribed by its family traits. But because ammonoids were such a fast-evolving order, there were deviations from the norm. These variable species—known as heteromorphs—are members of Ancyloceratina, a suborder where each exoskeleton may be a distinct variation on a common theme.

## Ancyloceratina

While planispiral homomorphs declined in diversity during the latter part of the Mesozoic, Ancyloceratina thrived, evolving into various bizarre genera. Their inconsistent shapes successfully filled specific ecological niches, and they became one of the stylemarks of pre-K-Pg extinction ammonoid fauna. Some atypical species were so widely distributed throughout Earth's seas that heteromorphs like *Scaphites hippocrepis* have been identified as a Cretaceous Period index fossil. As their dominance grew throughout the late Mesozoic, other Ancyloceratina—like baculites, nostoceratids, and additional scaphites—were adopted as time markers.

This category of unorthodox, irregularly spiraled suborder of *Ammonoidea* emerged in the Triassic (some sources may refer to it as an order), but their shells didn't radiate into shapes resembling modern sculpture until well into the Cretaceous. When you look at a

***Macroscaphites yvani*** **(Puzos, 1831)**
Order: Ammonitida. Appearance and lifespan: Early Cretaceous, 129.4 to 125 MYA. Location: Alpes-de-Haute-Provence, Provence-Alpes-Côte d'Azur, France. Size: 17 cm.

***Didymoceras nebrascense*** **(Meek & Hayden, 1856)**
Order: Ammonitida. Appearance and lifespan: Late Cretaceous, 83.5 to 70.6 MYA. Location: Bearpaw Shale, Rosebud County, Montana, United States. Size: 11 cm.

heteromorph genus like the boxy *Nipponites*, the spiral-hooked *Macroscaphites*, or the paper clip–like *Spiroxybeloceras*, you have to wonder how it maneuvered through the seas. Paleontologists have studied the outlandish Ancyloceratina and have deduced that this animal lived with the aperture directed toward the surface of the water, and the sculptural spiral of the conch unfurled accordingly.

This sundry ammonite suborder radiated to become one of the most distinct components of Cretaceous marine faunas. Members of Ancyloceratina had a thinner shell that coiled in some fashion other than a single-plain spiral. (The uncoiled anomaly among this group of abnormally conched cephalopods was the straight-shelled baculites.) There were Scaphitidae conchs that resembled the number 6 or 9. *Australiceras* and *Crioceratites* had tests that unraveled as they aged. These shapes were logical progressions, but then you had shells that were completely different. *Diplomoceras* had the contour of a dystopian trombone. The crazy-coiled genus had a conch that helixed into a loose, hooked tip. "Wrap a pipe cleaner around a

pencil and then tug it off and stretch it a bit. You can pull it into some crazy spiral shapes. And no two of them will look the same. Well, a *Didymoceras* looks kind of like that," explained Tony Lindgren, a collector of heteromorphs.

The open shells of these hooked cephalopods created a lot of drag, and their buoyancy was questionable. Propelling these nonhydrodynamic shapes made Ancyloceratina slow swimmers, and they were likely hungry from the effort. Helically coiled and orthocone Ammonoidea plausibly lived a benthic existence or clung to a reef and feasted accordingly. Analysis of stomach contents provides supporting evidence that these heteromorphs consumed a bouillabaisse of clams, molluscs, crinoids, worms, crustaceans, and the like.

A unique breed of Ammonoidea, Ancyloceratina probably did not share the efficient buoyancy control so prominent in other orders. Like a human with a space pack, this strange-looking ammonite could propel itself upward, apparently to scare aquatic pursuers. Also, this group could lurch its asymmetrical shell with an erratic

*Scaphites whitfieldi* **Cobban, 1951**
*Baculites* **Lamarck, 1799**
Order: Ammonitida. Appearance and lifespan: Late Cretaceous, 86.3 to 83.6 MYA. Location: Sandoval County, New Mexico, United States. Size: largest is 3 cm.

**Ammonite Bite:** One of the most common themes in heteromorph shell design is that of the conch spiraled on a single horizontal plane augmented by a hook-shaped living chamber. This bowed conch structure developed independently in at least five different lineages of ammonites during the Cretaceous, confirming its effective body plan.

technique, which provided survival advantage. "There was a lot of variety among cephalopods in the Cretaceous," confirmed exotic genus specialist Joseph Huston. "Heteromorphs were more like a small octopus rather than a nautilus."

## The Western Interior Seaway

Some 100 million years ago, Ancyloceratina were thriving, and, strikingly, North America was actually two landmasses. Divided by the Western Interior Seaway, the continent we know today was flooded with a massive body of water, which ran from the present-day Arctic Ocean to the southern gulf. The seaway stretched wide from central Utah to Illinois. Many states and territories, including Kansas, Colorado, Wyoming, and Texas, were underwater. Sources record that this sea, in its prime, was more than 3,200 kilometers (2,000 miles) long, 970 kilometers (600 miles) wide, and 760 meters (2,500 feet) deep. Along the shores, rich arrays of dinosaurs journeyed through the landscape.

An exceptional assortment of marine reptiles prowled the briny waters. Distinct species of ammonoids were part of communities that gathered around small, gassy elevated mounds on the seafloor. These hydrocarbon seeps extended above the anoxic bottom waters. Each created its own self-sustaining habitat that featured various combinations of ammonites, echinoids, sponges, new suborders of stalked crinoids, corals, and other invertebrate marine life. These sea geysers were a unique oasis in the barren seabed. "The hydrocarbon seeps around the Black Hills are a

A Late Cretaceous map of North America displaying the vastness of the Western Interior Seaway at its prime.

Image credit: Scott D. Sampson, Mark A. Loewen, Andrew A. Farke, Eric M. Roberts, Catherine A. Forster, Joshua A. Smith, Alan L. Titus, CC BY 2.5, via Wikimedia Commons.

to a rich diversity of marine life, including cephalopods, bivalves, gastropods, scaphopods, annelids, crinoids, echinoids, ophiuroids, crustaceans, bryozoans, corals, sponges, dinoflagellates, algae, fish, and reptiles. Each vent system seems to have hosted distinct life forms, with no two exactly alike, perhaps related to the maturity of each individual seep. Within some of these ancient ecosystems are families, genera, and species that have never before been found in the Western Interior marine environments of the Late Cretaceous."

In these nutrient-rich environments, the ribbed heteromorphs of the family Nostoceratidae unraveled like a corkscrew, dropping a U-shaped body chamber bend before the aperture. Diplomoceratidae were still a circular group, but their spiral became a wide-open whorl—a throwback reminiscent of some nascent Paleozoic Ammonoidea. But perhaps the most eccentric ammonite in the neighborhood was the one with the hooked helix conch known as *Didymoceras*. This special genus of heteromorphs evolved rapidly in the readily changing seaway, and area strata is identified by specific examples of these species.

Life thrived in and along the WIS for close to 60 million years. The slow uplift of the continent, accompanied by climate change, led to the disappearance of the channel toward the end of the Mesozoic. Water flowed toward the poles and became icebergs. "Throughout most of its history, this epeiric sea (i.e., a sea covering the interior of a continent) connected the Arctic Ocean (or Boreal Ocean) with the Gulf of Mexico (or Tethys Sea) and was joined to the North Atlantic Ocean, possibly episodically, via the Hudson Seaway," noted Dr. Joshua Slattery of the department of physics at the University of North Florida. "Sedimentation in the Western Interior Foreland Basin forms a thick and complex mosaic of interfingering marine and terrestrial deposits recording successive transgressions and regressions of the Western Interior Seaway."

treasure trove of fauna. The reason for so much diversity is probably related to the fact [that] the seeps provided a hard substrate in an otherwise vast expanse of muddy sea bottom. This led to the diversification and preservation of fauna that otherwise would have gone unnoticed," recognized a prestigious US research team skippered by WIS cephalopod experts Neal Larson and Neil Landman of the American Museum of Natural History. "These multifarious ecosystems served as a host

## We Dig Ammonites: US Heteromorph Locales

The variations created by growing landforms shifted the marine sedimentation increasingly eastward, creating hundreds of Cretaceous fossil locations throughout the lower forty-eight states. The Bearpaw and Pierre shales are laced with all types of Late Cretaceous seaway cephalopods. Among the other locales in the West where *Didymoceras* and other seaway ammonites have been discovered include the Mancos Shale and Williams Fork Formation in Colorado. There is also the Sego Sandstone in Utah, the Mesaverde and Rock River preservations in Wyoming, and the Lewis Shale in New Mexico. In the middle of the country, look for fine accumulations of Cretaceous cephalopods at the Bergstrom Formation in Texas, as well as the Ozan Formation and Annona Chalk of Arkansas. On the East Coast, explore the Mount Laurel preservation in Delaware and the Wenonah conservations in New Jersey.

"There have been a lot of papers published about different types of ammonites that are available in different horizons and different zones. By reading some of those papers, I was able to target some of the areas where *Didymoceras* were initially found. That information was accurate," shared geologist Steve Haire of Terrestrial Treasures. "There is a very fine stratum throughout which has all kinds of ammonites. But the *Didymoceras* is restricted to certain horizons. If you leave those areas, you won't find any Didys, for sure. Where we look is on private property. The landowners let us return because we respect the location." Exactly where to look and what the rules apply differ by location, so check with the local rockhounds when you plan your adventure.

Heteromorph specimens are typically found in outcrops and around creek bed concretions. At some sites, you will recognize earthen nodules

*Didymoceras cheyenense* (Meek & Hayden, 1856)
Order: Ammonitida. Appearance and lifespan: Late Cretaceous, approximately 101 to 66.0 MYA. Location: Pierre Shale, Pennington County, South Dakota, United States. Size: 20 cm.

**Ammonite Bite:** Clams are typically bottom dwellers. When you find preserved clams, look higher up on the formation to find ammonites.

sticking out of the embankments. Many of the loosely coiled species as well as straight-shelled baculites tend to be in sections, segmented by wave and current action or predation prior to fossilization. When you find a fragment, dig around it for the additional pieces. Perhaps you

**Acrioceras tabarelli (Astier, 1851)**
Order: Ammonitida.

**Phylloceras Suess, 1865**
Order: Phylloceratida. Appearance and lifespan: Early Cretaceous, 130, to 125.45 MYA. Location: Addar section, Agadir-Ida-Ou-Tanane Prefecture, Souss-Massa Region, Morocco. Size: scaphite is 8 cm.

can Frankenstein together a complete specimen. "If you spend some time prepping *Didymoceras*, you learn some tricks about how to work with the shape," offered Haire. "Some people think they're very difficult, but if you work with them long enough, you get the hang of it."

## Working with What You've Got

Logically, the heteromorph's center of buoyancy was the pivot point around which the shell rotated. The soft-bodied creature was agile—and may have only partially filled the living chamber. As the animal moved about its awkward conch, the ammonite's center of mass would shift the orientation of the asymmetrically shaped conch. When alarmed, the soft-bodied life form might turn the opening of its menacing shell away from the predator, retreating and shimmying back into the recesses of its protective covering. The thorny, helix conch and awkward movement likely confused any attacking predator. "These animals have been dismissed as

oddballs and mistakes," observed Kathleen Ritterbush, associate professor of geology and geophysics at the University of Utah. "But it is actually a perfectly executed plan, a coil of balance."

Some heteromorph genera had simple variations. Scaphites had shells that marginally departed from being a perfect spiral, except for the archetypal family overbite of the outermost whorl. With the genus *Tropaeum*, the last coil of the test separated from the spiral. The body section on *Ancyloceras* twisted underneath the main spiral in the shape of the letter *J*. In some species, the spiral section tilted to the side, resembling a hook with a jaunty beret. Just like adorable little babies may turn into bent old people, heteromorphs were not born with their shells spiraling in different directions. Genera such as *Scaphites*, *Macroscaphites*, *Nostoceras*, *Polyptychoceras*, and *Ancyloceras* all began life as cute, spiraled ammonitella, which transformed as the animal developed.

Juvenile heteromorphs were known to cluster away from shore. In some areas along the WIS, invertebrates were partial to methane seeps as well as the more traditional, fauna-rich environments. The symmetrical, planispiral juveniles of *Macroscaphites* and *Pravitoceras* were swimmers that could cover great distances before they entered their preliminary mating stage. Heteromorph shell divergence developed as ammonoids approached their reproductive phase. At this juncture, they were particular about their habitats. Scaphitidae preferred shallow shelves, where the speed limitations of its 9-shaped conch allowed it to hide safely, and there was plenty of plankton for sustenance. Ammonites like *Ancyloceras*, *Audouliceras*, and *Nostoceras*, which morphed more profoundly, opted for open waters. Their preferred environment and diet, like the creature itself, changed as the animal went through its life cycle.

Late in the Cretaceous, the Ancyloceratina was likely the dominant suborder in many areas of

A controversial ammonoid theory asserts that some hooked heteromorphs genera like *Didymoceras* attached themselves to seaweed as they matured. They developed a sort of prehensile shell which had the ability to shape itself around seaweed and other ocean flora, similar to modern-day seahorses.

***Didymoceras cheyennense* (Meek & Hayden, 1856)**
Order: Ammonitida. Appearance and lifespan: Late Cretaceous, 84.9 to 70.6 MYA. Location: *Didymoceras cheyennense* ammonoid zone, Pierre Shale, Carter County, Montana, United States. Size: 12 cm.

The heteromorph ammonite **Nostoceras** resembles a hook with a beret.

***Nostoceras malagasyense* (Collignon, 1971)**
Order: Ammonitida. Appearance and lifespan: Late Cretaceous, 83.6 to 72.1 MYA. Location: Belo sur Tsiribihinia, Madagascar. Size: 11 cm.

Image credit: Courtesy of Neal L. Larson.

the world's seas. Cephalopod families with more traditional planispiral shells had already become regionalized or even endemic to specific biota. This left many niches for hooked *Scaphites* genera to fill. They radiated and migrated, developing a considerably wider distribution in contrast to other pre-extinction Ammonoidea.

***Australiceras jacki* (Etheridge, 1880)** trio.
Order: Ammonitida. Appearance and lifespan: Early Cretaceous, 125 to 113 MYA. Location: Blackdown Formation, Walsh River, North Queensland, Australia. Size: 6 cm.

Carrier snail with cephalopod segment.

***Xenophora* Fischer von Waldheim, 1807**
Superorder: Hypsogastropoda
***Nostoceras* Hyatt, 1894**
Order: Ammonitida. Appearance and lifespan: Late Cretaceous, 72.1 to 66 MYA. Location: Big Brook, Monmouth County, New Jersey. Size: Nostoceras, 10 cm.

The quadrate shape of *Nipponites* likely gave the ammonite negative buoyancy.

***Nipponites mirabilis* Shimizu, 1935**
Order: Ammonitida. Appearance and lifespan: Late Cretaceous, 93.5 to 89.3 MYA. Location: Obata locality, Sorachi Subprefecture, Hokkaidō Prefecture, Japan. Size: 7 cm.

## Truth Is Stranger Than Fiction

Toward the end of their reign, Ammonoidea radiated more drastic heteromorphs. Perhaps examples of evolution's humor, these creatures radically departed from traditional ammonoids and their J-shaped family members. *Ptychoceras* hooked like a saxophone. The inexplicable *Nipponites* looked more like a three-dimensional box of connected U-shaped conduits than it did a living animal. *Nostoceras* combined different coiling modes: it started off with a helix but then grew a U-shaped loop underneath. Some were simply open spirals, while others resembled modern-day snails. *Didymoceras* looked like a Dr. Seuss hat, spiraling out with a curved tip for hanging. "Some of the heteromorphs matured into really crazy shapes," observed Huston. "The *Nipponites* may be considered one of the more eccentric creatures that has spent time on our planet."

Like homomorphs, heteromorphs varied in size. *Ptychoceras* were no longer than a finger, while straight-shelled *Baculites* and the switchbacked *Diplomoceras* approached six feet (1.7 meters) in length. Perhaps heteromorphs would have been the future of ammonites. Who knows what extraordinary creatures would have evolved if Ammonoidea had not been another victim of the Cretaceous–Tertiary elimination.

## We Dig Ammonites: Hokkaido, Japan

Collecting ammonoids on the Japanese island of Hokkaido may be a magical experience. It is possible to find more than five hundred species of Ammonoidea, and you can collect them while on a rafting expedition. Kanagawa prefecture on this northernmost large island of the Japanese Archipelago lets you float down the Mukawa River and collect fossils along the way. Several rafting

***Yokoyamoceras jimboi*** **(Matsumoto, 1955)**
Order: Ammonitida. Appearance and lifespan: Cretaceous, 86.3 to 83.6 MYA.
Location: Katsurazawa, Mikasa, Hokkaido, Japan. Size: 8 cm.

*Clioscaphites* **Cobban, 1951**
Order: Ammonitida. Appearance and lifespan: Late Creataceous, 89.8 to 86.3 MYA. Location: Hokkaidō Prefecture, Japan. Size: 3.5 cm.

companies offer packages. Inform company operators about your ammonite *yen* and you're on the way. This is one of those extraordinary places where you may step out of your boat and find Late Cretaceous cephalopods. Hidden in the landscape may be *Cheloniceras*, a deeply grooved evolute genus. Also on the scene may be *Colombiceras*, which had strong aesthetic ribs that wrapped around its narrow keel. *Gaudryceras* was smooth and evolute—a quintessential ammonite.

The fabulous preservation that retains all this Cretaceous sea life is known as the Yezo Group. It is part of a series of Late Cretaceous formations that extend some 1,500 kilometers (932 miles), spreading through Hokkaido, Japan, and Sakhalin, Russia. In the "Land of the Rising Sun," it is exposed on a roughly north-south trending belt from Cape Sōya in the north to Urakawa in the south. Municipalities in this area—including Nakagawa, Mikasa, and Mukawa—abound in fossils. "The city of Mikasa was under the sea 100 million

**Ammonite Bite:** Throughout Japan's forty-seven prefectures, the largest amount and variety of ammonite fossils have been found in Hokkaido.

years ago," observed the metropolitan museum's chief researcher, Daisuke Aiba. The Mikasa City Museum exhibits more than six hundred ammonites, among other extraordinary fossils. It is the finest cephalopod museum in the country.

Ammonoidea found on Hokkaido happen to be among the largest in the country. In the far north, it's possible to find iridescent ammonites, while in the central region, heteromorphs are more common. On rare occasions across the island, titanic ammonites are lifted out of the streams. *Yezoceras Elegans*, a cephalopod with conical spiral shell, was discovered by a river near the town of Haboro.

Hokkaido is indeed a cephalopod paradise, and while there are extraordinary heteromorphs here, those locations are private and treasured. The *ojisan* ("old uncles") know the pockets where coveted specimens such as *Yezoceras*, *Pravitoceras*, and *Nipponites* are located. Relationships with the knowing seniors are built, not bought, and a translator is a necessity unless you are fluent in the local dialect.

## The Legalities of Fossil Prospecting in Japan

From the northern tip of Hokkaido to the southern end of Kyushu, Japan, you can find more than 2,800 fossil-collecting sites. Yet the Japanese word for "fossil"—化石 / *Kaseki*—is not referred to in government documents. As a result, information on the export of fossils for personal use from Japan is difficult to confirm. "I don't think there are any rules," noted a source intimately familiar with the Japanese fossil market.

Search the term "fossil export Japan" online and what is returned is a report on how many watches have sold on the nation's Home Shopping Network. Add the word "cephalopod," and the results reveal import/export sales on extant squid and octopus used for consumption. The most relevant information can be discovered on the fossil forums. Those in the know say that the terms to use are "cleaned, dirtless rock" and "geological sample."

The most reliable details refer to mailing fossils out of the country, but the information can be extrapolated for customs. "Japanese fossils aren't prohibited from international shipping, but sometimes troubles show up at the post office," noted David in Japan. "The reason is quite stupid: When shipping something from Japan, you have to declare the content of your package . . . As fossils aren't part of the Japan Post nomenclature—contrary to gems / rocks or animals—you will be asked by the clerk to describe what a fossil is."

Fossils are the remains or impressions of prehistoric organisms. They are replicated in petrified form or as a mold or cast in stone. You have the information, make your pitch, describe your finds as "cleaned/dirtless rock" and/or "geological sample." The clerk will use their best judgment to decide if the fossil is classified as a rock (permineralized) or animal (wildlife). Japan permits rocks but not animals to leave the country.

Many display ammonites are prepared for presentation. These pyritized ammonites were shined using a brass brush.

***Pleuroceras solare* Phillips, 1829**
Order: Ammonitida. Appearance and lifespan: Early Jurassic, 189.6 to 183 MYA. Location: Nuremberg area, Middle Franconia, Bavaria, Germany. Size: 6 cm.

# Ammonites Are Sacred

People around the planet have been compelled to imbue *Ammonoidea* with metaphysical powers. Since the early days of the human race, a tapestry of myth, legend, and folklore has developed around these coveted cephalopods. These endearing shells have become part of custom, fable, and tradition to a cosmopolitan coterie of cultures. Tales from ancient times tell the story of how these sea creatures were named in honor of the divine and all-powerful Egyptian god Amun. The king of the deities is most readily identified by the curled rams' horns sprouting from his head, creating the visual link that resulted in these cephalopods being named *Hammonis cornu* ("the horns of Amun"). This moniker was attached to Ammonoidea as early as 77 CE, when Roman author and naturalist Pliny the Elder identified them in an early version of an encyclopedia. Rumor even has it that they may have been given such a powerful name because this respected member of the intelligentsia, slept with a pyritized ammonite from Ethiopia under his pillow. Pliny the Elder deemed that having a golden cephalopod near one's brain "could make the dreams and aspirations of the possessor come true," cited Paul D. Taylor, scientific associate and merit researcher emeritus at the Natural History Museum in London.

Shaligrams are considered sacred by millions of people.

***Blanfordiceras wallichi* (Gray, 1832)**
Order: Ammonitida. Appearance and lifespan: Late Jurassic, 152.1 to 145 MYA. Location: Muktinath, Mustang District, Gandaki Pradesh, Nepal. Size: 6 cm.

## Shaligrams = Blessed Black Cephalopod Concretions

For more than two thousand years, the black limestone concretions of the Late Jurassic cephalopods found in the Mustang District of Himalayan Nepal have been sacred to the Hindu people. Devotees call them *shaligrams* (or *saligrams*), and they are believed to be imbued with special powers. "True saligrams are found only in the valley of the Gandaki River, close to the village of Salagrama and the town of Muktinath in Nepal," elaborated fossil folklore specialist Taylor. "They come from a Late Jurassic–Early Cretaceous deposit called the Saligram Formation, which is correlated with the better-known Spiti Shale."

Hindu and Buddhist traditions throughout Nepal and the Indian, attest that the Kali Gandaki is the only place on Earth to host the five elements from which all material things in the universe are made. In this remote paradise, the majesty of Earth, air, sky, water, and space are apparent and clearly delight the senses. The fire element is unique. It is present in the form of a flame fueled by naturally occurring gas outflow (the tectonic plates are still active). This thermal seam heats nearby springs, giving these pools the appearance of water on fire. Unique to the elevation, it is said no other water source with burning-gas locations are known to co-exist on our planet. Beyond the five elements, what makes these locally discovered ammonites special are their supposed metaphysical powers. It is indeed miraculous that these 150-million-year-old sea creatures might be found some 3,900 meters (12,795 feet) in the sky in the landlocked, snow-heavy Himalayas.

"Mentioned in Sanskrit texts dating back to the second century BC, saligrams are kept in temples, monasteries and households as natural symbols of Vishnu," observed Taylor. "Water in which saligrams have been bathed is drunk daily. In addition, saligrams are used in marriages, funerals, and housewarmings. If a dying person sips water in which a saligram has been steeped, it is believed that they will be freed from all sins and will reach the heavenly abode of Vishnu."

Officially known as *shaligrama shila* (शालिग्राम शिला) ammonites collected from the riverbed or escarpments of the Kali Gandaki gorge and nearby

tributaries are venerated by more than a billion admirers. These Jurassic cephalopods are revered by followers of Hinduism's Vaishnava, Shaiva, Shakti, and Smarta doctrines and devotees of India's Jain Dharma. It is appreciated by adherents to the religions of Tibet and the Himalayas: Buddhist and Bön (an ancient creed with temples that may feature the design of a left-handed swastika.). All those believers consider shaligrams to be special because of the long and complex saga that resulted in a curse being laid on Lord Vishnu, thus turning the deity into a stone. The theology goes that the supreme being is embodied in the Kali Gandaki ammonite geodes, making them inherently sacred. Conveniently, they are easy to carry and are freely available to all who seek them. Beyond the confidence of always having a blessing in your pocket, these sacred concretions are traditionally incorporated into home altars as a figurative representation of Lord Vishnu. "As unbounded forms of the divine, they are unlike such man-made objects as idols, images or statues [murtis])" explained Holly Walters, a cultural anthropologist specializing in South Asian Studies. "And because they require no rites of incantation to consecrate them, each shaligram is the deity. To be in the presence of a shaligram is to be in the constant presence of the gods themselves."

## We Dig Ammonites: The Himalayas

Several varieties of fossils occur high in the Himalayas, but shaligram cephalopods are found only along the Kali Gandaki River and its tributaries in the Mustang District. These revered specimens thrived during a relatively brief time, from 150.8 to 145 million years ago. These special stone shells are found in the shale outcrops of the Annapurna massif, which towers above the river and reaches elevations up to 5,000 meters (16,400 feet). It is an excursion for the hale and hearty.

Here, you will find several species of the genus *Blanfordiceras*, a strongly ribbed, evolute ammonite that has radiated from the superfamily *Perisphinctoidea*. Struts bifurcate on the outer flanks and reach onto the venter. It is easy to confuse with the genus *Berriasella*, which has a more rounded whorl section. Another that looks similar is the

***Ataxioceras* Fontannes, 1880**
Order: Ammonitida. Appearance and lifespan: Late Jurassic, 157.3 to 152.1 MYA. Location: Gräfenberg area, Bavaria, Germany. Size: 5 cm.

evolute, rough-ribbed *Corongoceras*. You may locate the familiar *Lytoceras*, which is evolute; with whorls that are somewhat quadrate; and is covered with some type of curious texture, likely crinkled growth lines, riblets, or fine striations running the width of the flank. Look for the evolute *Ataxioceras*, which has a shell with thirty-two to thirty-six sharp, spaced ribs per whorl. Rising from a large, wide, flattened center, the struts trifurcate about mid-flank. The involute *Haplophylloceras* had spokes like a tractor tire.

## Sri Muktinath

Should you choose to take the time and make the effort to visit this extraordinary, exotic remote mountainside in the Himalayas, you will find that shaligrams are easy to retrieve from the riverbeds of the Gandaki tributaries. Here, on top of the world, ammonite adventurers may also find foreigners gathering shaligrams for their religious pilgrimage to the central shrine of the Vishnu temple complex of Sri Muktinath. This unpretentious sanctuary is perched on a mountainside near Nepal's border with Tibet and is treasured by Vaishnava, Hindus, and Buddhists. Here, the pious receive *darshan*—the beholding of a deity, revered person, or sacred object. The experience is considered reciprocal and results in the human viewer receiving a blessing. Based on this philosophy, the darshan blessing occurs when one finds a shaligram.

The fossil continues to radiate this power long after the Nepali ammonite has been placed on an altar or passed on to the next generation, as a treasure embraced by the family.

Because the Gandaki riverbed is just downslope, Sri Muktinath has been the principal site for the veneration of local Jurassic ammonite geodes since Lord Vishnu is said to have turned to stone. Wander around outside the temple, and you will likely meet a pilgrim proudly showing off their blessed, black, spiraled fossil concretions. "Shaligrams. Very old. Very sacred," they may say reverentially, thus extending an olive branch of friendship by speaking in English.

This mountain paradise is located in the eastern segment of the lengthy Alpine-Himalayan orogenic belt, a 15,000-kilometer (9,300-mile) range of

## The Legalities of Fossil Prospecting in Nepal

The Federal Democratic Republic of Nepal wants you to visit, enjoy yourself, boost the economy and take home souvenirs. The government has embraced policies that work in ammonite collectors' favor: (1) they encourage tourism revenue, and (2) the country is looking to increase exports to reduce the trade deficit.

Feel like you are being blessed when you purchase a shaligram from locals who set up shopping blankets covered in shaligrams and yak wool accessories and other locally crafted treasures. Commercial transactions in Nepal are all about boosting the country's economy and improving the quality of life for its inhabitants.

**Ammonite Bite:** Rumor has it that there are tribes on the Upper Sepik River in New Guinea who continue to use ammonites as charms that offer energy to aid with hunting and farming.

crustal squeezing. These cordilleras are the result of the closure of the Tethys Ocean and subsequent abrasion between the northward-moving African, Arabian, and Indian plates as they butted the Eurasian Plate. This action began in the Late Mesozoic Era and is still building the Indian subcontinent and the Tibetan plateau. Today, this Alpide Belt is the second most seismically active stretch on the planet.

*Discoceras* **Barrande, 1867**
Order: Nautiloidea. Appearance and lifespan: Late Ordovician, 455 to 449.2 MYA. Location: Pupiao Formation Shidian, Baoshan, Yunnan Province, China. Size: 1.5 cm.

In a world before the media, it's not hard to imagine that this baculite chamber resembled a buffalo.

Image credit: Sosanna, CC BY-SA 4.0, via Wikimedia Commons.

## The Middle Kingdom

In the Land of the Red Dragon, the earliest writings about fossilized bones are Chinese records that date back to the fourth century BCE. It wasn't long afterward that these same writings acknowledged ammonoids. Folklore in the Far East also likens these cephalopods to coiled rams' horns, so they referred to *Ammonoidea* as *Jiǎoshí* (角石) hornstones. They were described in the herbal compendium *Pen Tshao Thu Ching* around 250 BCE. A rough translation of the text would be, "The stone-serpent appears in rocks which are found beside the rivers flowing into the southern seas. Its shape is like a coiled snake with no head or tail-tip. Inside it is empty. Its color is reddish-purple. The best ones are those which coil to the left. It also looks like the spiral shell of a conch. We do not know what animal it was which was thus changed into stone."

## The Legend of Blackfoot Buffalo Stones

Several Native American tribes, including the Indigenous Blackfoot people of Southern Alberta,

Canada, have long appreciated the intrinsic power of ammonites and baculites. The First Nations cultures prize cephalopod chambers as buffalo stones and have been sharing tales of their metaphysical properties for more than a millennium.

"When the chambers break off, they naturally have this bison shape. It's amazing. There's a head, then the back and four little legs," illuminated Evelyn Siegfried, curator of Aboriginal studies at Canada's Royal Saskatchewan Museum.

Upon discovery in the wild, these cephalopods' chambers have likely been kissed by the wind, and the rain has stripped the sheen from the nacre, yet that does not affect their unique attributes. Generations of Indigenous Canadians have passed along the folktale about how a buffalo stone helped an entire tribe of starving people. "This stone is magic," said Troy Knowlton, one of nearly forty thousand members of the Blackfeet Nation in Canada and the United States.

The origin of the tale of the buffalo stone dates back more than a thousand years. A Blackfoot tribe that camped in the plains around Alberta was suffering through a particularly long frigid winter season.

The snow was so deep and the storms so severe that the people could not move about in search of buffalo. This was indeed a crisis because the mighty animal was vital for survival on the Canadian plains. Not only did the creature supply plenty of food but its hide could also be made into blankets and clothing and used for teepee material. The tribe tried to meet their needs by traversing the river bottoms and the ravines. They killed deer and elk and other small game, but this prey was soon hunted or scared away. The Blackfeet were facing hunger. In desperation, the outfit prayed to their maker for relief so the tribe would not starve. That night, one of the women had a most special dream that a spirit visited her and revealed, "The Creator has heard your prayers, and he sees your struggle, he has sent me, a spirit to give you a gift." The divine one gave the woman a detailed set of instructions, all focused on an *iniskim* (also known as buffalo stone):

Part 1: Find the amulet. The dream mystic described landmarks near where the mystical ammonite chamber was hidden. "You'll hear this song coming from the stone," the specter noted, and shared the melody.

Part 2: Once the buffalo stone was retrieved, the tribe was to perform a specific ceremony of ritual, chant, and song to honor the iniskim.

Part 3: After the ritual, two events would follow, and then the buffalo would arrive.

When she awoke in the morning, the woman told her husband of the dream spirit that had visited her while she slept. He listened thoughtfully and told his wife, "Prepare, and go find the stone."

Realizing his role in the spiritual ritual, the husband gathered the people of the camp together and shared, "Creator has heard our prayers, and he's answered us!" With the blessing of her people, the woman set out through the valley. In time, she noticed the landmarks indicated by the dream

spirit. In the distance, she heard the song. She followed the tune, and the volume increased as she grew closer. The music was emanating from a preserved ammonoid chamber that looked very much like a buffalo. She picked up the precious fossil and returned to camp. Presenting the ammonite amulet to her husband, she stated, "This is the gift. Tonight, we shall perform the ceremony as the spirit instructed."

After dusk, the tribe elders gathered at the lodge, and the woman presented the buffalo stone to the family. The leaders learned the song of the talisman. The ritual was then spread throughout the tribe, and together they prayed that they would not starve.

The specter told the clan that they would have two signs that would secure the promise of buffalo, and it was necessary that the group act accordingly. The divine foretold of a severe storm that would descend from the north. The tribe was instructed to bring in their personal belongings, tie up their dogs, and tether down their teepees. The second sign was a buffalo that would wander through their camp later that night. The tribe was not to harm this buffalo. Should they follow these instructions, when the sun rose, their hearts would be glad.

The tribe performed a heartfelt ceremony singing and dancing in honor of the iniskim. They were hungry and hopeful, and they went to bed for another long cold winter's night. In the darkness, the words of the divine came to pass. The forecasted storm whooshed through their settlement, and those who hadn't prepared had their teepees blown over and their personal effects whisked about. Then the dogs barked and howled, and when the people looked out, they could see a buffalo wandering through the camp. They knew it was a message from the creator, and they did not harm the creature.

When the sun rose, all was calm. Snow drifts were strewn across the valley. Just outside the

"Buffalo hunt, chase" from George Catlin's 1845 *North American Indian Portfolio: Hunting Scenes and Amusements of the Rocky Mountains and Prairies of America.*

Image credit: George Catlin, public domain, via Wikimedia Commons.

camp, a small herd of buffalo were trapped chest deep in snow. The people of the camp were able to go out and terminate the creatures. The kill provided food, shelter, and clothing throughout the rest of the winter.

"The buffalo stone was used as a power of spirit to call the animal. We had to hunt buffalo, and the hunt had to be successful, or you wouldn't eat. So we had diverse ways to assist us, and the buffalo stone played an important role," explained John Murray, the tribal historic preservation officer for the Blackfeet Nation in Montana.

According to Wild West field biologist and zoologist Joseph Grinnel, buffalo stones were "found on the prairie, and the person who succeeds in obtaining one is regarded as very fortunate. Sometimes a man who is riding along on the prairie will hear a peculiar faint chirp, such as a little bird might utter. The sound he knows is made by a buffalo rock . . . If it is found, there is great rejoicing."

Since the time this legend was first shared, buffalo stone ammonoid segments have been a powerful symbol that is sacred to the Blackfeet

Nation. To this day, cephalopod segments resembling buffalo are prized talismans believed to bring assistance, luck, and prosperity along one's twenty-first-century journey. "There are ceremonies where the rocks are blessed and painted. The women dance with the stones, while the men sing the songs," shared Knowlton, who is a professional Bearpaw Shale ammonite ammolite hunter. "It's an amulet that provides more eyes to see upcoming danger."

Several versions of this tale circulate among the First Nations people. Some interpretations of the story talk about the events actually happening. Other versions describe it as a vision that came to a woman in a dream. Some have the buffalo stone speaking; others have a visit from a midnight spirit.

## Medicine Bag Cephalopods

The Indigenous people of the Americas are a resourceful group. Not only were the individual chambers of ammonites considered sacred, the shell's ammolite nacre was also deemed to have medicinal qualities. This biogenic gemstone was regularly described as part of the contents of medicine bundles: wrapped collections of spiritually significant items. A packet might include feathers, pieces of ammolite, bones, buffalo stones, animal skins, roots, clay pipes, and (later) tobacco. Every article in the medicine bundle had spiritual significance and called for a special song whenever its owner unpacked the talismans. The parcel itself was traditionally made of hide and beads and had its own prescribed chants and myths.

Transfer of a remedy bundle from one generation to the next was a hallowed process that involved a solemn ceremony. The new owner had to learn the significance of all the objects in the packet, details of visions to which they owed their origins, and the

*Placenticeras* **Meek 1876**
Order: Ammonitida. Appearance and lifespan: Late Cretaceous, 83.5 to 70.6 MYA. Location: Saint Mary River, Lethbridge, Alberta, Canada. Size: approximately 35 cm.

**Ammonite Bite:** Medicine men of the Navajo Nations and other Plains tribes carried ammonite amulets. These cephalopods were known as *wanisugna*, and the shell was described as "life within the seed, seed within the shell."

incantations that released their influence. When the amulets in the parcel were useful, feasts may have been offered in honor of the medicine bag. Blackfeet historian Bob Scriver acquired several remedy bundles in the early twentieth century. In many of the packages were what he calls buffalo calling rocks, which were often wrapped in the animal's fur. According to legend, the cephalopod chamber calling stone was placed inside a rawhide *parfleche* and hung outside the lodge on its own small tripod to call in buffalo.

Group of Siksika (Blackfeet) men and one woman singing in front of a tribal tepee.

Image credit: Rodman Wanamaker (1863–1928), copyright claimant, public domain, via Wikimedia Commons.

## Ammonites as Healing Tools

Much of Europe was underwater while Ammonoidea swam in the planet's seas, and today these cephalopods may be found throughout Europe. Colloquially known around the Old World as dragonstones, crampstones, serpentstones, and snakestones, they have been part of European culture and ritual since the days of the Saxon clans and the Teutonic tribes. These feudal fraternaties imbued ammonoids with the inherent powers and purposes that met their medieval needs and put the coiled concretions to practical use. Those on the Western Isles of Scotland referred to their ammonoids as crampstones and used them to heal cattle. To facilitate the cure, the fossil was steeped in water for some hours. The medicinal liquid was then used to wash and swaddle the painful parts of the cow.

Ammonites were put to similar uses by farmers in southwest England. In the Celtic nation of Cornwall, coiled cephalopods called snakestones were part of potions used to treat simple maladies. "Beasts which are stung, being given to drink of the water wherein this stone has been soaked, will therethrough recover," wrote Richard Carew in the turn-of-the-seventeenth-century *Survey of Cornwall*. Additionally, Anglo-Saxon England is reputed to have used ammonites to hinder vision loss, promote fertility, and cure impotence. It was also held as a sympathetic medicine to prevent and cure snakebites.

In the Harz highlands of northern Germany, the homesteaders used *Ammonoidea* as medicine to treat livestock. Folklore suggests that the settlers employed a fossil shaped like a ram's horn when cows lost their milk or passed blood. Referred to as drake and / or dragonstone, the practice was to put the fossil into the teat bucket to stimulate the cow's lactation. "They believed that adding a dragonstone to the milk pail would help ensure the return of milk to cows that had stopped producing," explained Taylor of London's Natural History Museum.

In a smattering of southern villages along the Jurassic Coast, ammonites were thought to have been mischievous fairies that morphed into snakes, which then coiled up and turned to stone. The concept of cephalopods as petrified serpents has also been traced to areas of continental Europe as well. Legends of ammonite dragonstones from southwest Germany were reinforced when twenty-first-century anthropologists unearthed a water-logged grave. The preservation held what is believed to be an affluent (and perhaps mystical) sixth-century Celtic woman. She had been buried with her valuables. Among the gold jewelry and furs were a fossilized ammonite and sea urchin, which has the locals theorizing that she may have been a priestess.

## Snakestones

The early inhabitants of British Isles habitually developed folklore and legend around *Ammonoidea*. In several of the kingdom's medieval realms, beliefs linked cephalopods to snakes. "Stories about snakestones came primarily from two places where ammonites are very common and easy to find: Whitby in Yorkshire and Keynsham in Somerset," noted Taylor.

One of the earliest known legends about eels and ammonites is from the parish of Keynsham in

The people of Whitby have been carving snake heads on ammonites since the Middle Ages.

*Dactylioceras commune* **Sowerby, 1815**
Order: Ammonitida. Appearance and lifespan: Early Jurassic, 182 to 175.6 MYA. Location: Whitby Formation, Yorkshire Coast, England, United Kingdom. Size: 7 cm.

Somerset, southwestern England. In Back in the days of pagan Britain, round about ye olde fifth century, lived the princess Keyne, one of the twelve daughters of the Welsh king Brychan of Brycheiniog. An adventurous and deeply religious woman disinclined toward marriage, the noble lass was given land on the River Avon to make her home. It was a mixed blessing because the marshy lowlands slithered with snakes. Folktale goes that Lady Keyne prayed, and the serpents turned to stone. Ammonite concretions were displayed as evidence of her success, but the details of this miracle have been obscured with time. It's said that Lady Keyne spent many of her adult years traveling through the region spreading the word of Christianity. The

*Hildoceras* is named in honor of Saint Hilda of Whitby.

**Hildoceras bifrons Bruguiere 1792**
Order: Ammonitida. Appearance and lifespan: Early Jurassic, 182.7 to 174 MYA. Location: Larzac Region, France. Size: 5 cm.

incredibly rich is today). Traditionally, to receive sainthood required several achievements:

1. The person must epitomize Christian virtues.
2. They must have a "reputation for holiness," at least after their death.
3. They must have performed a miracle.

Just as dinosaurs can inspire legends about dragons, a bevy of spiral ammonite geodes discovered by god-fearing people can be an inspiration for sainthood. In northeast England, the town of Whitby has patron Saint Hild, also known as Hilde or Hilda (614–680), who was canonized thanks, at least in part, to ammonites.

In medieval Europe, no one could explain cephalopod concretions, so curious folklore arose to justify their existence and purpose. Known in pre-Christian England as serpentstones and snakestones, these concretions typically hid coiled cephalopods like *Dactylioceras* or *Hildoceras.*

Born in Britannia, a province of the Roman Empire along the North Sea, young Hilda was of noble blood. Growing up in the seventh century, the young lady was given unparalleled exposure to Saxon diplomacy. Political intrigues swirled about her early life, and she developed the savoir faire to charm aristocrats and to curtsy for kings. So savvy was Lady Hilda that she was offered the opportunity to wield power—an extraordinary prospect for a noblewoman of the Early Middle Ages. The caveat was that she could only do so as a nun. Through family influence, Sister Hild became the liaison to a Roman consortium sent by the Pope. A meeting with Anglo-Saxon King of Northumbria led to Hild being appointed as the founding abbess of the new church. In 657 CE, on a bluff along the Yorkshire Coast, Sister Hild and the missionaries established the Christian monastery of Streoneshalh (later to be rechristened Whitby Abbey). She had the responsibility of building the

church even granted her sainthood because of her efforts. There is an endearing legacy about Saint Keyne that takes place around a holy well named in her honor in the village of Cornwall. Since the Middle Ages, believers have maintained that when a married couple approach the fount of Saint Keyne, the spouse that takes the first drink will have the upper hand in the marriage.

## The Ordination of Ammonites

In medieval Britain, sainthood could be awarded like gold medals. There was no formal canonization process. Local deification was ordained by the bishop of the diocese. Martyrdom was a goal that could hypothetically be achieved (like becoming

cloister that would transform the pagan populous into Christians.

As captured in the classic novel *Wuthering Heights*, the wild heathered knolls of the Yorkshire moors were there to be tamed. Sister Hild had selected a hillock with a magnificent view, but the land chosen for Streoneshalh was plagued with snakes. Undaunted by the slinking creatures, Sister Hild first attempted to clear the land using prayer. "In early Christian times, snakes had a bad reputation and were associated with the Devil, so it was important to clear the area before a sacred building could be established," elaborated Taylor.

Unlike the fabled success of Saint Keyne, Sister Hild's pleas went unanswered. She tried various Old Testament methods to clear the land of the vermin. She and others attempted fires, floods, horses, hoes—yet the snakes continued to make the bluff their home. Frustrated, the consortia beseeched their maker and wondered whether this reptile-infested soil could ever accommodate humans. Supplication finally yielded divine inspiration; Abbess Hild tried the snap of a whip as she uttered another round of prayers. According to folklore, eyewitnesses and faithful Christians watched as the snakes curled up into balls and rolled off the cliffs, never to return. When ammonite concretions were found at the base of the bluffs, the devoted believed the snakes had turned to stone.

Stories circulated throughout the kingdoms about how the Abbess Hild had defeated the devil disguised as a serpent. It was hailed as a miracle. The lady superior received sainthood, thanks in no small part to the ammonites at the base of the ridge. The streamlined evolute genus *Hildoceras* is named in her honor. You can identify these fossils by their picket ribbing with a narrow channel accenting their inner flanks, and a fluted keel.

During her lifetime, Saint Hild became the patron saint of learning and culture, with fourteen medieval churches bearing her name. Saint Hild's College at the University of Oxford was founded in her honor in 1893. Its mission is to foster intellectual and personal achievement.

**Ammonite Bite:** The sound and motion of the snap of a whip terrifies snakes and leaves them disoriented.

The obvious inconsistency to this story is that Whitby snakestones do not have heads. Saint Cuthbert had the answer. Apparently, a spell issued by the Christian martyr decapitated the asps. "The versions of the legend that include Saint Cuthbert help explain the headless state of Whitby's snakestones," shared Taylor. "He is said to have cast a powerful beheading curse on all of the snakes." (FYI, this seventh-century priest was also said to have had a rosary made of fossilized crinoid segments.)

## Stranger Than Fiction?

The legend of Saint Hild and the ammonites sounds like a tall tale, yet her story is documented in *The Ecclesiastical History of the English People*, written in 731 by the Benedictine monk, Bede. (He also chronicled the English version of the Anglo-Saxons' conversion to Christianity.) Scottish novelist Sir Walter Scott also penned a verse about the sainted noblewoman in the poem "Marmion, a Tale of Flodden Field": "How of a thousand snakes, each one was changed into a coil of stone, when holy Hild prayed."

These prophetic events occurred around the Middle Ages hamlet of Whitby. Located along the North Sea moors of Yorkshire, it is notable for its natural riches, the success of its herring and whaling fleets, and tourism. Through the centuries, Whitby has stayed true to Saint Hild and her ammonites. The town coat of arms depicts "three coiled serpents against azure waves." As part of local lore, artisans have been carving snakeheads into local ammonites since Abbess Hild's sainthood. Some of the early cut and polished specimens from Whitby sailed as far away as Norway, indicating that they were precious enough to be traded and transported. Perhaps spoil from a Viking age pillage, these serpent stone specimens spread the legend of Saint Hild, and ultimately the appreciation of ammonites.

The Whitby coat of arms features three ammonites on a shield of blue waves.

Image credit: Coat of Arms of Whitby, Yorkshire, England, granted under Crown authority by the College of Arms to Whitby Urban District Council 1935, assigned Whitby Town Council.

**Ammonite Bite:** It was in Whitby during the eighteenth century, mariner James Cook learned seamanship. He later became a captain in the British Royal Navy and navigated three voyages to the Pacific.

## The Lias Group

Scriptures from the Middle Ages recall that snakestones littered the bluffs of Whitby like shells on a beach. Little could those medieval minds have envisioned that these ammonite-strewn cliffs and coves would one day be a world-class paleontology location. At that moment in history, they would likely have had a difficult enough time imagining fossils. It took nearly one thousand years after Saint Hild for humankind to recognize the paleontological richness of this North Sea coast. Perhaps the first written reference to the region's fossiliferous riches was published in the book *Britannia* in 1586. Author and cartographer William Camden observed, "If you break them you find within stony serpents, wreathed up in circles, but eternally without heads."

Ammonoids, paleontology, and the concept that land was once underwater did not take hold in early modern Europe until the Age of Reason. In 1676, the first description of a dinosaur bone appeared in English literature. The specimen was discovered embedded in the Taynton Formation at the Stonesfield limestone quarry, Oxfordshire, United Kingdom. The illustration that accompanied the report is the first known published image of a dinosaur bone, what we identify as *Megalosaurus*, and it became the first genus to be described and named. French paleontologist Georges Cuvier evaluated and contrasted living animals and fossilized creatures, establishing the fields of comparative anatomy and paleontology. His study made its way across the English Channel, allowing those erudite British men of science to chronicle paleontological findings in the first known English-language scientific journal, *Philosophical Transactions of the Royal Society*. They compiled a report on a "Skeleton of an allegator found in the Allom Rock near Whitby, January 3, 1758."

Interest in paleontology spread, and Whitby became a go-to spot for the geologically inclined. When entrepreneurial Alum quarry workers uncovered large skeletons, they sold them as curiosities. These skeletons were subsequently transported to other parts of Britain, with no information recorded for study. To protect these rare and historic marvels, the Whitby Literary

*Psiloceras (Caloceras) johnstoni* **Rakús, 1999**
Order: Ammonitida. Appearance and lifespan: Early Jurassic, 201.3 to 196.5 MYA. Location: Hettangian, Somerset, England, United Kingdom. Size: 12 cm.

and Philosophical Society was formed in 1823. Their hall became the repository of the precious paleontological items that eroded out of the area hills. This coveted horde became the foundation of the awe-inspiring Whitby Museum, the ideal place to admire area ammonites and marine reptiles. Around that time, geologists professionally acknowledged the fossiliferous limestone and named the entire biota the Lias Formation. It encompasses some of the most important geological formations in Britain, including the Dinosaur Coast.

The Lias Group consists mainly of fossil-friendly clays, mudstones, and limestones deposited between 205 and 180 million years ago. The Late Triassic to Early Jurassic English outcrop extends in a continuous band from the south coast of Dorset to Yorkshire in the northeast, with some outlying

*Hildoceras bifrons* **Bruguiere 1792**
geode.
Order: Ammonitida. Appearance and lifespan: Early Jurassic, 183 to 176 MYA. Location: Whitby, Yorkshire, England, United Kingdom. Size: 11 cm.

areas in Somerset and South Wales. A favorite collecting spot for fossil hunters and paleontologists, savvy diggers may discover a petrified dinosaur footprint, the jaw of an ancient crocodilian, or a juvenile ichthyosaur. When you hunt for fossils there, you will no doubt find the area's most famous fossil—a snakestones, also known as an ammonite concretion—most likely a *Dactylioceras, Hildoceras* or *Eleganticeras.* These serpentstones are abundant in the upper Lias rocks, particularly along the Whitby Scar that lies between the tidelines. The locals suggest looking along the high water mark. Preservations are typically discovered on the foreshore between the high and low water marks. Also look at the base of the cliffs, where the snakes are said to still be recoiling from Saint Hild's wrath.

## We Dig Ammonites: Whitby

Strategically positioned off the coast of Europe, the United Kingdom is exposed to the North Atlantic Ocean on three sides. Some serious waves pummel the Yorkshire coast, exposing wonderful ammonites and other Jurassic treasures on a daily basis. Near Whitby, you'll find fossils along the beach and near the cliff face—but be careful because parts of the escarpment loosen and fall to the ground below. (British Isles folklore recollects how experienced fossil gatherer Mary Anning lost a dog when a piece of the precipice of her hometown cliffs came tumbling down.) When you search for fossils along the British coast, it's an excellent idea to head out with a professional. Around Whitby, you'll get a detailed lesson in identifying ammonite nodules and how to split the concretions to reveal the fossils inside. Your local expert knows the tide tables and will make sure you get back to dry land before the sea comes rushing back in, pinning unknowing beachcombers against the cliffs. These tours often provide rock hammers, chisels, and pith helmets. The stories shared by the guides are quite memorable and worth the price of admission. Your leader, likely speaking with a jaunty British accent, shall share local tales, ghost stories, and mysteries of this ancient seaside village.

Depending on the tides, a professionally conducted fossil-hunting trip should offer you a tour of the town, with a stroll around the ruins of Whitby Abbey as well as a hike down to the small, sandy cove of Saltwick Bay. This ammonite-rich location

Whitby Coast, North Yorkshire, England, United Kingdom.

Image credit: Andrew Bone from Weymouth, England, CC BY 2.0, via Wikimedia Commons.

is about a kilometer east of downtown. The long flats of creased intertidal rock provide a feast for wading birds that dine on crabs and other fauna. An experienced collector, your resident mentor will guide you to the safest and most prolific spots to find preservations. Ammonite concretions may reveal themselves with bits of the shell peeking out around the edges of the knobble. "They look like a fossil hamburger when you find them—a rock in the bread bun," chuckled area geologist and paleobiologist Byron Blessed of Natural Wonders Ltd. "You usually find them with the ammonite partially sticking out."

Patience is a virtue when it comes to fossil hunting because it can take time to acclimate your eyes and your senses to the treasures hidden in the surroundings. Much of successful preservation excavation comes down to noticing shapes, structures, and patterns that stand out from the landscape. Treasure nodules tend to be a distinctly different color from the surrounding stone. Depending on the type of preservation, the node could be slightly darker or lighter and the texture and weight will be different. Look for anomalies that don't match the matrix and recurring patterns in the rocks.

Once you have identified a concretion to your liking, tap it along the centerline, and it may magically split to reveal an adpression specimen. The positive side of the rock has the actual fossil, whereas the negative is the impression. With North Sea geodes, the negative can display just as much detail as the positive. Hiding inside you may find pleasingly spiraled *Ovaticeras*, *Pseudolioceras*, or *Eleganticeras*. Confirm your find with a local expert because all three were members of the Harpoceratinae family and had involute compressed

**Ammonite Bite:** Examine the cliff face about one meter (a yard) above beach level because reptile bones can occasionally be found along this tier.

Whitby ammonite geodes are locally known as cannonballs.

(*top*) *Dactylioceras* **Hyatt, 1867**
Order: Ammonitida. Appearance and lifespan: Early Jurassic, 182 to 175.6 MYA. Location: Saltwick Nab, Whitby, Scarborough, North Yorkshire, England, United Kingdom. Size: 8 cm.

(*bottom*) *Eleganticeras elegantulum* **Young and Bird, 1828**
Order: Ammonitida. Appearance and lifespan: Early Jurassic, 183 to 182 MYA. Location: Whitby, Scarborough, North Yorkshire, England, United Kingdom. Size: largest is 3.5 cm.

shells with strong keels and similar ribbing. *Eleganticeras* is often found in extremely rounded nodules known as Whitby cannonballs.

If you want to dig without a guide, access the find zone by Whitby's East Pier. It is essential that you know the tide times and be sure to return before the waves get too close to the cliffs. The safest and most prolific collecting stretches are about three to four hours after high tide. Algae and kelp create sticky areas; be aware because it is easy to slip or get stuck, especially after a rain.

The beaches and cliffs are often covered in standing water, and the routes are uneven, so a sturdy pair of shoes with well-gripping soles is imperative. An easy fix when in rainy old England is to do like the British do: pick up an inexpensive pair of Wellington boots. You will need a chisel and a geological hammer for striking open the geodes you find. Safety goggles are best worn when cracking into concretions. Record your collections: time, place, depth, description of the specimen, and location. When you are transporting them home, wrap your newfound treasures well using clear wrap and place them into plastic containers to keep them intact during transport. Have them ready for airport security to inspect. The United Kingdom is unique because it is pleased when you take commercial grade fossils home with you. Otherwise, these specimens would just fall into the sea.

## Jet?

As you browse the shops around Whitby, you will notice many Victorian antiques and jewelry trimmed in a semiprecious black stone christened jet for its midnight anthracite color. This mineraloid is a form of fossilized wood from the Early Jurassic Monkey Puzzle Tree (also known as Araucariaceae) that has morphed under extreme pressure. A type of lignite—the lowest rank of coal—it is indeed a gemstone, broadly speaking. Whitby proudly boasts that the world's finest quality jet is found under the cliffs or in seams of shale along its coastline. If you are looking for preserved wood or local amber, check under and around patches of seaweed as pieces often get tangled in the kelp. Many of the bits of black "rock" you find are actually coal washed up by the sea. Check your discovery by rubbing it on a sandstone pebble or a piece of sandpaper. If the mark is black, it is coal; if it is brown, you have found a piece of Whitby jet. The

Whitby jet with *Psiloceras* ammonite impression.

**Psiloceras Hyatt, 1867**
Order: Ammonitida. Appearance and lifespan: Early Jurassic, 180 MYA.  Location: *Falciferum* Zone, Jet Rock Series, Upper Lias, Whitby, Yorkshire, England, United Kingdom. Size: 3.5 cm.

tideline is also a good place to discover little pieces of fossilized wood.

Whitby is by far the easiest locale for fossil hunting, but if you want to expand your digging adventure and your species finds, the following sites in north Yorkshire, United Kingdom, are some of the neighboring coastal cliff formations rich in ammonite fossils:

## South

Robin Hood's Bay is a postcard-picturesque, historic hillside fishing village an easy 10 kilometers (6 miles) south of Whitby. Be advised, however, that the area yields but a few fossils during the summer months, and it is often only after winter storms that finds are made. Residents of this quaint seven-hundred-year-old hamlet suggest that you access the beach from the footpath to the left of New Road. The trail takes you up the hill on the north side of the village, past the picnic area. Focus on finding ammonites around the exposed shales at low tide.

Embedded belemnite fossil discovered along the shore of Saltwick Bay, near Whitby, Scarborough, North Yorkshire, England, United Kingdom.

**Belemnites Lamarck, 1799**
Order: Ammonitida. Appearance and lifespan: Middle Jurassic, 163.5 to 157.3 MYA. Location: Saltwick Bay, Scarborough, North Yorkshire, England, United Kingdom. Size: approximately 10 cm.

Image credit: Michael Jagger, CC BY-SA 2.0, via Wikimedia Commons.

***Pleuroceras*** **Hyatt, 1867**
Order: Ammonitida. Appearance and lifespan: Early Jurassic, 189.6 to 183 MYA. Location: Péault, Vendée, Pays de la Loire, France. Size: largest is 5.5 cm.

Here, you can discover *Platypleuroceras* or *Acanthopleuroceras*, but good luck telling them apart. They are both evolute, ribbed spiral ammonites with a pronounced spine. Should you find *Tropidoceras*, you'll see that it had fewer coils, and the genus gets wider as the shell grows. *Androgynoceras* had very round whorls and thick, wide-spaced ribs that wrap across the venter. If you want more than Ammonoidea, there are crinoids, belemnites, and bivalves to be discovered. Locations like Filey, Cayton Bay, Speeton, and Gristhorpe Bay all have cephalopods, and the occasional remains of *Ichthyosaurus*. You can see dinosaur tracks in Scalby.

## North

Just beyond the Whitby Golf Course is Sandsend Ness. The cove at the north end features shale banks laden with easy-to-harvest belemnites. Kettleness is 10 kilometers (6.4 miles) up the coast and has a remote, strategically located north-facing inlet where there are regular finds of ammonites and jet. Next to it is Runswick Bay, which has rock pools to explore. Ammonites are commonly found hidden among the tidal rubble. Port Mulgrave is a former ironstone export harbor with a steep path to the beach. Here, you will not find many people looking for the plentiful cephalopod and rarer reptile fossils. On the top end of the Dinosaur Coast is Staithes, the eighteenth-century home of British Royal Navy Captain James Cook. It's a good location for finding iron pyrite.

The rules for collecting along these remarkable area in northeastern of England are similar to those of most fossil sites, when you discover something unique, the Dinosaur Coast Code expects you to report it so they can determine if it is scientifically important. The bright side is that, unlike most other countries, you get to keep what you find, even if scientists ask to borrow it for study.

## Easy Pickings

Whether you're a beginner rockhound on holiday or an experienced collector, Whitby is a wonderful place to find ammonites. Your memorable days on its beach will be peppered with enthusiastic comments, and you are likely to come away with as many ammonite geodes as you can carry. "I never tire of finding fossils," observed our guide, Whitney from Whitby. As if it were a movie moment, she picked up an oddly shaped rock and gently tapped it with her hammer and chisel. It cracked in half, and some of the stone shards fell to the ground. Inside was a small *Hildoceras*. "There's just something so magical about revealing a creature that lived so deep in the past that has never been seen before. It's like stepping into a time warp to another planet." The backpacks were already chock full of better specimens, so the ammonite named in honor of Saint Hild was left for someone else to discover in the hope of inspiring their lifelong passion for ammonites.

***Jeletzkytes* Riccardi, 1983**
Order: Ammonitida. Appearance and lifespan: Late Cretaceous, 83.5 to 70.6 MYA. Location: Pierre Shale, Glendive, Montana, United States. Size: 4 cm.

# They Arrived After One Extinction and Expired from Another

In the Late Cambrian Period, the seas were a seemingly endless expanse of open ocean with few sizable creatures. Animal life consisted largely of benthic bottom dwellers: those early arthropods—the trilobites—explored the midnight zones, accompanied by eely conodonts, curious crinoids, and early molluscs. Maybe it was something in the water, because species diversity blossomed. One of those taxa was Nautiloidea—the senior subclass of shelled cephalopods. Its diversification into numerous ecological niches ushered in a dramatic change for life on Earth.

Evolutionary conditions on our planet were pristine for much of the Ordovician Period until geomorphic events instigated an extinction incident. Some 443 million years ago, a series of cataclysmic occurrences—reportedly activated by the rise of North America's Appalachian Mountains—led to the Ordovician–Silurian extinction. The freshly uplifted rocks apparently sucked carbon dioxide out of the atmosphere, triggering extreme global cooling and an ice age commenced. Vast glaciation locked up enormous amounts of water in a frozen field that covered extensive parts of the southern landmass. The episode obliterated an estimated 85 percent of all species; quantities and varieties of corals, shelled brachiopods, conodonts, and trilobites were wiped out. In terms of the percentage of genera that became extinct, the Ordovician annihilation is said to have resulted in the second largest disappearance of creatures in Earth's history.

Western Interior Seaway ammonite mortality plate.

*Moremanoceras straini* **Kennedy et al., 1988**
Order: Ammonitida. Appearance and lifespan: Late Cretaceous, 99.6 to 93.5 MYA. Location: Eagle Ford, Texas, United States. Size: largest is 3 cm.

Imagine a normal day in the semitropical Cretaceous Period about 66 million years ago. Dinosaurs were happily munching on woody cycads, shady ginkgoes, and leafy fern plants. Ammonoids were frolicking in the water column near the continental shelves and doing their best to evade larger predatory vertebrates like mosasaurs, plesiosaurs, sea turtles, large bony fish, and perhaps aggressive birds. Then, with no apparent warning or reason that Mesozoic life could possibly comprehend, a massive object—an asteroid wider than the Grand Canyon—came hurtling toward them at about 45,000 miles an hour. It blocked the sun like a solar eclipse, and then *KERPOW!* The Chicxulub asteroid hit with the power of several million nuclear weapons. It sent out a flash of energy that ignited landscapes beyond a radius of 1,450 kilometers (900 miles). The chaos of fire was compounded by earthquakes, storms of flaming pellets, and tsunamis. Today, the impact area sediments splay the area of the Yucatán Peninsula in Mexico. The devastation ravaged the Americas from North Dakota to Patagonia and as far east as New Jersey—perhaps it was only the water boundaries of the Atlantic Ocean and the Western Interior Seaway that stopped the current of chaos.

On the fateful day of the Cretaceous–Paleogene (K-Pg) extinction event, Earth was forever changed. "Coarse grain size, abundant subangular rip-up clasts, and imbricated debris in the basal portion indicate that the initial stages of deposition were rapid, turbulent, and high energy," observed paleontologist Robert DePalma and his research team. "Climbing ripples, pronounced grading of the deposit, water-escape structures, truncated flame structures, and steady vertical transition from upper- to lower-flow regime-flow structures provide additional proof that the accumulation of the sediment package was brief and episodic." The impact wedged invertebrate fossils into terrestrial animal burrows.

The fauna that rallied in the wake of this global cooling was able to adapt to relatively rapid changes in temperature and water chemistry. When conditions finally stabilized and the oceans reoxygenated, it became a veritable free-for-all as animals evolved to fill ecological niches left vacant by the extinction episode. Cephalopods of the straight-shelled order *Ellesmerocerida* were some of the lucky few sea creatures that survived the tumultuous times. They went on to yield several new orders and became a foundation for the proliferation of Ammonoidea.

Not quite 400 million years after ammonoids appeared in Earth's seas, they went extinct. The circumstances under which this cephalopod subclass left the planet were not all that different than how they arrived: it was the result of an extinction event.

The dense bone-colored ribbon marks the Cretaceous–Paleogene boundary. It contains 1,000 times more iridium than the upper and lower layers. This band has been found throughout the world in greater or lesser degrees.

Image credit: Zimbres, CC BY-SA 3.0, via Wikimedia Commons.

## What Happened?

After the torrent of terror was unleashed by the initial impact, the planet succumbed to a nightmarish sequence of events that devasted the global environment for decades. A dust cloud of carbon gases, vaporized rock, sulfur, iridium, and other toxins eventually settled across Earth's surface. The planet was plunged into a lingering impact winter, with the debris in the atmosphere decreasing the amount of sunlight able to reach our planet's surface. This combination of occurrences inevitably proved fatal to most shelled cephalopods (primarily ammonites and belemnites), but it spared some of those that scavenged along the sea floor. The Chicxulub incident minimized plant photosynthesis and created a transient episode of surface water acidification that temporarily curtailed dinoflagellate replication. (In paleontology, the term "temporary" can mean hundreds to tens of thousands of years.) Plankton feeders like the ammonitella

and other juvenile sea life starved to death. These shelled cephalopods disappeared, as did bivalves, gastropods, and the marine reptile superfamilies that dieted on those organisms, Bye-bye, mosasaurs, and so on up the food chain. When the toxic dust cleared, 76 percent of species—including ammonites and terrestrial dinosaurs—had disappeared. "Ammonoids and belemnites had small offspring and high metabolism: this means they had little reserves but a high need of energy," concluded a research team led by Christian Klug and Amane Tajika.

"Scientists spend a lot of time trying to figure out the cause of ammonite extinction," elaborated cephalopod authority Neil Landman of the American Museum of Natural History. "It is generally linked to the disappearance of calcareous plankton following the bolide impact due to an ephemeral episode of ocean surface acidification.

Either the ammonites perished because they fed on these animals or because they themselves were part of the calcareous community during their early life history. On the other hand, no one spends much time thinking about the consequences of ammonite extinction, either the effects on the animals the ammonites ate, or on the animals that ate the ammonites. It is still an area to explore."

More than 66 million years later, paleontologists are still interpreting the results of the asteroid impact. A fascinating remnant still under investigation and interpretation is a thin, unusually dense, silvery-white layer present at Cretaceous–Paleogene boundaries throughout the world. This clay contains elevated levels of the chemical element iridium. The most corrosion-resistant of metals, it is considerably more abundant in meteorites than in Earth's crust. This unique stratum records secrets to the global impact of this mega asteroid strike. The K–Pg saga was the most recent—but certainly not the only—naturally inducted elimination crisis to devastate Earth in the 540 million years since life has existed on the planet.

## Did Ammonoidea Die Immediately?

How long the once mighty subclass lived after the Chicxulub event is something that keeps paleontologists up at night. Biostratigraphy indicates that during the last several million years of the Mesozoic Era, Ammonoidea was decreasing in quantity and species variety. Only four suborders—Ammonitina, Ancyloceratina, Lytoceratina, and Phylloceratina—were still swimming during the last 500,000 years of the Cretaceous Period. Radiations from these groups had generated six superfamilies and thirty-one genera, but half of all their species had varied and endemic geographic distribution. Some ammonite types were as rare as

*Sphenodiscus* **Meek, 1871**
Order: Ammonitida. Appearance and lifespan: Late Cretaceous, 70.6 to 66 MYA. Location: Pierre Shale, South Dakota, United States. Size: 27 cm.

panda bears and flourished in a select few sites. What lived where depended on the marine conditions. Bays, lagoons, and seaways were the Late Cretaceous ammonoid settings of choice, and they proliferated throughout much of continental Europe and the Atlantic coastal plains of North America. "The varied morphologies represented by this diverse group of organisms reflect mode and life and to some extent environmental conditions," observed Richard Batt, American geology educator. "Although most ammonites apparently ranged from shallow nearshore to deeper mid-basinal habitats in the Western Interior Seaway (maximum depth about 300 meters [984 feet]), the distributions of certain morphotypes suggest depth restriction. For example, compressed, disc-shaped, and heavily nodose forms generally indicate bottoms shallower than about 50 meters [164 feet]. Some morphotypes appear to reflect a nektobenthic and others a pelagic habitat. The presence of specific morphotypes may be used to indicate benthic and water-column oxygenation, respectively."

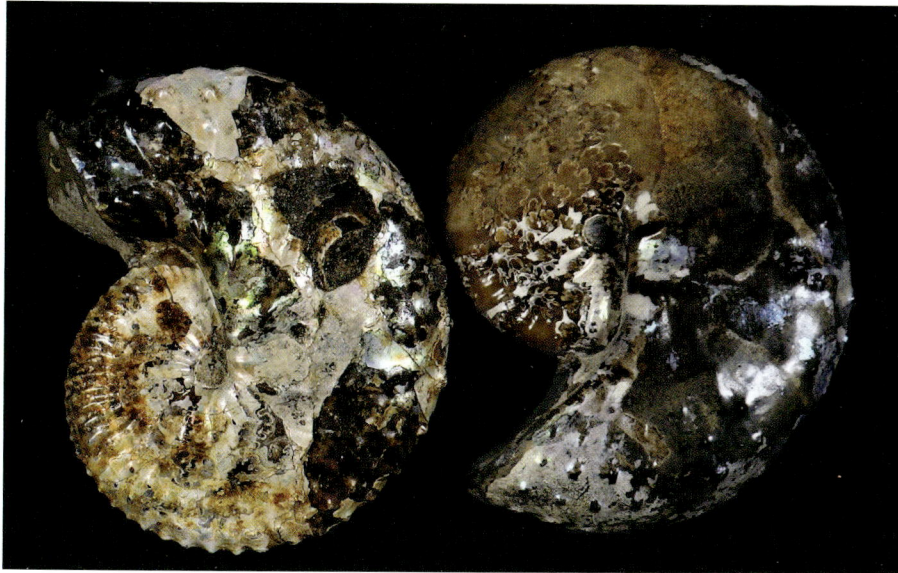

***Discoscaphites conradi*** (Morton, 1834)
Order: Ammonitida. Appearance and lifespan: Late Cretaceous, 70.6 to 66 MYA. Location: Hoploscaphities nicoletti Zone, Trail City Member, Fox Hills Formation, Corson County, South Dakota, United States. Size: 5 cm.

In these epipelagic waters, shell shapes varied more radically than at any time in *Ammonoidea* history. The compressed, disc shaped Sphenodiscidae was buoyant and free swimming. The many family members of the rapidly radiating 9-shaped heteromorph Scaphitidae and the trombone-styled Diplomoceratidae sometimes aggregated around seeps and reefs. Baculitidae were likely the most abundant, adaptable, and widespread members of the Ammonoidea subclass at the end of the Cretaceous Period. Their weighty, tapered shafts made them sluggish swimmers, so they often hovered in the water column at an inclined angle, snatching delectables as they swam by. In the colder waters of the boreal regions, the species that prevailed was the thick and hearty *Desmoceratidae*; largely involute, it had a deep umbilicus. A more sophisticated ammonite, it preferred feeding along the continental slopes and had developed a buccal apparatus that permitted it to consume a larger diet. The most notable innovation was a calcareous deposit at the end of its lower jaw, perhaps to pierce crustacean conchs and other carrion.

As you would suspect, extinction patterns varied according to latitude and distance from the Chicxulub impact. Not unlike other disasters, such as the eruption of Mount Vesuvius in 79 AD near Pompeii, the 1906 San Francisco earthquake, and the 1986 Chernobyl nuclear disaster, those further from the impact zone had more time before the consequences took effect. The greater the distance, the better the chance for the toxic byproducts to diffuse. "If the ammonites had not gone extinct, they might still well be tooting along," observed Landman. "The species in shallow water habitats like baculites might have suffered due to competition with teleost fish and predation from shell-crushing crustaceans, but deeper water forms like the desmoceratids might have retreated to the same niche occupied by nautilus—steep forereef slopes with depths up to several hundred meters."

## We Visit Ammonites: They Tried to Live on in Denmark

**Name: Stevns Klint**

**Location: Denmark**

**Age: Cretaceous**

**Date of UNESCO Inscription: 2014**

At Stevns Klint in Denmark, the extinction event is memorialized by a dark thin layer, known as *fiskeler*.

Image credit: Niels Elgaard Larsen, CC BY-SA 3.0, via Wikimedia Commons.

Stevns Klint is a UNESCO World Heritage Site, one of those remarkable stratigraphical and paleontological phenomena. It displays a prominent iridium layer, and it verifies that at least two types of ammonites survived beyond the K–Pg extinction event that wiped out an estimated 75 percent of all species on the planet. Paleontology and technology have revealed that at this Baltic Sea location, *Hoploscaphites constrictus* and *Baculites vertebralis* lived on, at least for a little bit.

The fossil record at Stevns Klint provides a breathtaking succession of three ecosystems. Stretching 17 kilometers (11 miles) along the coast on the Danish island of Zealand, this outstanding succession of three biotic assemblages provides context for the K-Pg boundary layer by exposing one of the most diverse end-Cretaceous marine fossil records discovered to date. The precipice provides evidence of the preimpact community; the collision; and what taxa became extinct, which survived, and the tempo of their recovery. The embalmed fossil record reveals the complete succession of fauna and microfauna, allowing us to better chart recovery after the mass extinction.

In addition, stratigraphers say that this is the most significant and readily accessible site to see the sedimentary record of the ash cloud formed by the asteroid's strike.

Located about an hour's drive south of Copenhagen, the millions of years captured in this preservation-rich rock face reveal evidence of the global effect of the Chicxulub asteroid that ended the Cretaceous Period. The amazing fossil record at Stevns Klint spans from the Mesozoic to the Cenozoic—a 10-million-year swath that straddled the Late Cretaceous and Early Paleocene. The cliffs share the story about the evolution of the enduring marine wildlife and how those creatures diversified into our modern-day sea dwellers. At this extraordinary site, the impact results are memorialized by a dark thin layer, regionally known as *fiskeler*. Called fish clay in English-speaking countries, this iridium-rich ribbon holds the details of the decades following the K–Pg extinction event. The detritus of the disaster is punctuated with dots of impact glass preserved below the atmospheric dust that settled across the entire planet. The powder in the air scattered onto the

***Hoploscaphites nebrascensis* (Owen, 1852)**
Order: Ammonitida. Appearance and lifespan: Late Cretaceous, 70.6 to 66 MYA. Location: Fox Hills Formation, Dewey County, South Dakota, United States. Size: 5 cm.

**Ammonite Bite:** The Bryozoa chalk in the cliff at Stevns Klimt is highly shock resistant to both conventional and nuclear weapons.

ocean floor, cementing itself atop a thick layer of algae. The fallout from the atmosphere formed the chalk sea bottom for the postapocalyptic world. The marine survivors that lived into the Paleogene Period are found above that level, in the limestone at the top of Stevns Klint. The fish clay is visible above the surface of the sea, but you cannot be sure of getting close to it. To see the details, use binoculars or augment the viewing experience with the telescopes at Højerup Old Church or from the beach below.

## The Survivors

On the Cenozoic side of the fiskeler layer are the remains of the holdover taxa—those that lived beyond the extinction event. Rare specimens of the heteromorph ammonites *Baculites vertebralis* and *Hoploscaphites constrictus* have been found embedded in situ at one of the postelimination Cerithium Limestone basins that cross the region. These species confirm data from Denmark and the Netherlands, which concluded that a few types of resilient ammonites may have survived up to 200,000 years after the Chicxulub elimination event.

Stevns Klint is not the only location with Ammonoidea survivors. Assemblages of boundary ammonites have been found in other high-latitude sites in Denmark and in the southeast Netherlands, Poland, Turkmenistan, northern Greenland, the Antarctic Circle, and even New Jersey in the United States. Scientists have hypothesized that, because these land locations were so far from the Chicxulub impact site, it took longer for the consequences of the event to take effect. Paleontologists believe there are other areas where ammonoid species briefly broke through the extinction barrier. Most of these yet-to-be-discovered locations are likely covered in permafrost.

*Maorites seymourianus* **Kilian and Reboul, 1909**
Order: Ammonitida. Appearance and lifespan: Late Cretaceous, 83.5 to 66 MYA. Location: Seymour Island, Weddell Sea, Antarctica. Size: 7 cm.

## Seymour Islands

One of the most complete Cretaceous sedimentary successions in the Southern Hemisphere occurs just outside the Antarctic Circle, where the Drake Passage meets the Weddell Sea. Known as the Seymour Islands, they are lonely little spots of land in a remote and rugged archipelago around the tip of the Antarctic Peninsula. Formations on the isles offer comprehensive biostratigraphic preservations from a time when the bottom of the world was considerably warmer and had dinosaurs. A unique environment with heavy rainfall and polar forests, it was once a lush preserve of conifers, cycads, and ferns. The Late Cretaceous fossil record conserved many endemic species as well as several genera that had gone extinct in other parts of the planet. In the seas, new ammonite, plesiosaur, and mosasaur species continued to appear—although there happened to be more disappearances than appearances. Conchs from the suborders Lytoceratina, Ammonitina, and Ancyloceratina are embedded in sediment with enhanced concentrations of iridium. Even more interesting is that there are ammonites above the transformation matrix that is presumed to be Antarctica's version of the K-Pg boundary (it is not clearly defined at the bottom of the world). Several specimens have been collected from horizons up to 13 meters (43 feet) above the accepted extinction border. Discoveries like these bolster the theory that *Ammonoidea* lived for a while after the Chicxulub asteroid event. Among those that appear to have made it through were a half dozen or so distinct species from the family Kossmaticeratidae. Antarctic area identifications include the regional genus *Maorites*, which had an involute shell with fine ribbing and nodes on the interior of the flank. A few members from the superfamily Desmoceratidae appear to have survived—their large round shells look like they could have devolved into snails, but they didn't. The involute ribbed ammonite *Pseudophyllites* is another Antarctic refugee that had already perished elsewhere. The traditional-looking *Anagaudryceras*; the trombone-like *Diplomoceras*; and the big, round, involute *Pachydiscus* are also believed to have survived into the Danian Period near the bottom of the earth.

**Ammonite Bite:** Seymour Island along with neighboring James Ross Island comprise the largest ice-free surface in Antarctica and are referred to as oases. These isles have extensive dry valleys of jagged exposed stone and minimal snowpack. Volcanic dust that settled into the ravines has minimized the brutal ice accumulations. Here, at the bottom of the earth, ice and land regularly erode, exposing new sections of the rock surface. Argentina maintains the Estación Marambio for research, as well as a gravel airstrip to monitor the islands. As an added bonus, seventeen varieties of penguin live throughout the archipelago.

Arctic ammonites were decidedly different from Antarctic ammonites.

**Surites Sazonov, 1951** nodule.
Order: Ammonitida. Appearance and lifespan: Early Cretaceous, 145 to 139.8 MYA. Location: Krasnoyarsk Krai, Arctic Siberia, Russia.
Size: 5.5 cm.

## More Ammonite Survivors

In the unlikely location of Monmouth County, New Jersey, ammonites have been discovered above the iridium layer. Uncovered by construction workers building bridge foundations, the Pinna Layer represents a time from tens to hundreds of years after the impact. The formation was identified by state park ranger and amateur paleontologist Ralph Johnson in 2003. Capturing a diverse nearshore marine environment, the Pinna Layer preserves at least 110 species of small sea life. In the mix were several continuing ammonite types with a variety of body styles, including the circular *Pachydiscus*, the hooked-chambered *Discoscaphites*, and the straight-shelled *Eubaculites*. Some of the cephalopods were discovered in an undisturbed bed of triangular fan clams that died jutting upward, indicating that this once thriving community may have been buried by pulses of mud-rich sediment, possibly associated with flooding from freshwater rivers.

It has been documented that several species of *Hoploscaphites* attempted Cenozoic Era breakthroughs in Atlantic Ocean passages in both hemispheres. Pockets of after-impact heteromorphs have been recognized at multiple South Atlantic locations around Chile and Argentina, not far from the Seymour Islands. Further north along the WIS, elegantly preserved Danian stage *Hoploscaphites* specimens have been identified in Colorado, Nebraska, South Dakota, and Montana.

*Arctocephalites* **Spath, 1928** also known as ***Cranocephalites costidensus* (Imlay)**
Order: Ammonitida. Appearance and lifespan: Middle Jurassic, 167.7 to 164.7 MYA. Location: Tuxedni Formation, Talkeetna Mountains, Alaska. Size: 5 cm.

## The Cenozoic Era

Although there were some laggards that attempted to save the subclass, the K-Pg extinction event essentially marked the end of Ammonoidea, four-legged terrestrial dinosaurs, and all land-dwelling vertebrates larger than a small dog, as well as the Mesozoic Era. After the devastation of the Chicxulub asteroid impact, it took the planet some thirty thousand years to right itself and continue on its evolutionary path. The new environment had mammals filling the niches left by the terrestrial reptiles. The dog-sized *Hyracotherium* grew into the horse, and Omomyidae was an early primate that swung from the trees. Warm-blooded creatures dominated the landscape, developing into what we know as cows, elephants, whales, bats, cats, apes, puppies, and a whole host of critters that birthed live young. Avian birds are the only surviving creatures descended from dinosaurs, and they in turn radiated the modern species of feathered avifauna.

In the seas, a bout of ocean acidification essentially annihilated the plankton, which in turn eliminated the ammonites as well as most other creatures feeding in the sunlit top waters. That impact ran down the food chain, purging mosasaurs, plesiosaurs, and other large swimming reptiles. Groups of bony fish and many other molluscs were lost to the devastating conditions.

Ammonites struggled to hang on, and they succeeded in some areas for a while. In most cases, fossils and other traces of animal life are abundant below the iridium layer and sparse, if at all, above the boundary, with few exceptions. Although *Ammonoidea* had survived and flourished after several mass extinction events throughout their nearly 400-million-year evolution, they finally perished in the wake of the end-of-Cretaceous Chicxulub bolide impact. "If they made it through this event like they did through other mass extinctions, why didn't they take off again?" questioned invertebrate paleontologist Peter Harries of the University of South Florida.

Noted ammonoid researcher Christian Klug of the University of Zürich explained, "There are two problems, which made ammonoids vulnerable: (1) they had rather high metabolic rates (like belemnites and cuttlefish), and (2) their hatchlings were minute (about 0.6 to 1.2 mm). The combination of these two factors likely meant their demise because the hatchlings did not get enough food and lacked reserves to survive longer phases without food."

The transitional nature of global biota does indeed implicate changes in climate chemistry as a significant factor in faunal turnover. In the oceans, the acid rain fallout from the asteroid impact proved to be a passing phase, but its short-term disruption was so severe that it has been linked to the disappearance of *Ammonoidea* and other animals that surface-fed on phytoplankton. Existence was easier for benthic creatures able to digest detritus from the seafloor, far away from the acidified surface waters. This is likely the reason *Nautiloidia* continues to exist.

**Prionocyclus hyatti (Stanton, 1894)**
Order: Ammonitida. Appearance and lifespan: Late Cretaceous, 93.5 to 89.3 MYA. Location: Sevier County, Utah, United States. Size: 6 cm.

## The Silver Lining

If you look for a bright side, the Chicxulub asteroid impact opened underground hydrothermal systems that became an oasis for the recovery of sea life. Some of the species that rejuvenated after the impact—various fish, sea urchins, algae, and sharks—were smaller due to the limited food supply. The most sizable air-breathing survivors of the event were semiaquatic ancestors of the crocodile, which could survive on detritus. On the surface of Earth, almost all predators larger than a footstool were gone. The surviving groups developed into the extinction gaps. The creatures that were spared adapted, modified, and thrived, laying the foundation for life as we know it today.

The view from an overlook near Hell Creek on Charles M. Russell National Wildlife Refuge, Montana, United States.

Image credit: Paula Gouse, U.S. Fish and Wildlife Service, public domain, via Wikimedia Commons.

## We Visit the Last Ammonites: Hell Creek

**Name: Tanis, Hell Creek Formation**

**Location: Southwestern North Dakota**

**Age: Cretaceous**

**U.S. National Natural Landmark Designation: 1966**

North America was still forming at the end of the Cretaceous Period. Much of the west central part of the landmass was flooded by the large shallow Western Interior Seaway, an intercontinental extension of the Atlantic Ocean that ran from the Arctic to the southern gulf. What we know as the Hell Creek Formation was a partly submerged cove at the northern end of the watercourse. Although the actual bolide strike was some 4,750 kilometers (2,952 miles) away from Tanis, Hell Creek and other locales shook as if they had been walloped by an 11.5-magnitude earthquake. The tremors generated seismic waves between 10–100 meters (33–328 feet) high, which rolled in all directions. The impact tsunami sped up the WIS, carrying sea, land, and freshwater life inland for dozens if not hundreds of miles. Those marine swells likely reached Hell Creek about 10 minutes after the asteroid crash. The fossil assemblages that have been found here preserve extraordinary

Cretaceous and lower Paleocene rock stretches across portions of the states of Montana, North Dakota, South Dakota, and Wyoming. The Tanis Lagerstätte site enshrouds the freak circumstances caused by the impact event.

## We Dig Big Vertebrates

Hell Creek is one of the most famous dinosaur-bearing formations in the world, which is why most people come here for vertebrate fossils. (Ammonite fans tend to go to the Fox Hills outcrops.) Fossil-hunting privileges are leased out, and by far the easiest way to go is a guided dig. There are options online, and you can sign on for a couple of days or a couple of weeks. Your paleontology group may visit colorfully named private locations like Tooth Draw Quarry and Longbone Ranch; each site is a pocket of something special. Locals who have been collecting in this area since they were smaller than a *Dakotaraptor* ulna guide the paleo crews to a seemingly innocuous ancient riverbed. And there, hidden amid the sand and well-rounded pebbles of this lag deposit, are fossil prizes like gar fish scales, crocodile teeth and scutes, turtle shell fragments, and occasional river-tumbled dinosaur bones and teeth. Treasures are plentiful. It is not unusual to find an edge-worn T-Rex tooth, a raptor radius, an *Edmontosaurus* jaw section, part of a *Triceratops* rib and / or teeth, hadrosaur vertebrae, as well as a trove of other preserved body bits. Check your dig tour contracts: if finds are rare and scientifically valuable, they're often kept by the quarry master to be shared for research purposes. Donated finds tend to be dinosaur parts, like the peaked bony back plate from an armored *Ankylosaurus* or a hollow bone with predation marks from a feathered *Acheroraptor*.

The Hell Creek dig spots are pebbly sand blended with clay, which makes it easy to excavate. Tools to have on hand include a small shovel,

**Placenticeras intercalare**
Order: Ammonitida. Appearance and lifespan: Late Cretaceous, 83.6 to 72.1 MYA. Location: Bear Paw Shale, Rosebud County, Montana, United States. Size: 2 cm.

devastation situations. Ammonites have been found by burned tree trunks. Fish were unnaturally stacked one on top of the other. Conifer branches impaled turtles and entwined mosasaur bones. Nearby, partially exposed carcasses of struggling dinosaurs imply sudden death by catastrophe. This place of distinctive devastation is in the southwestern North Dakota section of the expansive Hell Creek Formation. Coursing for roughly 700 kilometers (435 miles) from east to west and up to 100 meters (328 feet) thick, Hell Creek's expansive Late

***Collignoniceras woollgari* (Mantell, 1822)**
Order: Ammonitida. Appearance and lifespan: Late Cretaceous, 94.3 to 89.3 MYA. Location: Carlile Shale, Newcastle, Weston County, Wyoming, United States. Size: 4 cm.

scalpel, screwdriver, paintbrush, and glue. To view an impressive collection of area fossils, go to the Museum of the Rockies, in Bozeman, Montana, or the Rocky Mountain Dinosaur Resource Center, Woodland Park, Colorado.

**Ammonite Bite:** If you're looking for K-Pg boundary ammonites from the Western Interior Seaway, find the areas of Hell Creek that overlay the Fox Hills Formation. This locale is a fine place to find some of the last surviving ammonite families such as *Sphenodiscus* and *Discoscaphites*.

A pride of ownership piece may be a naturally occurring conglomerate of different species.

---

*Euhoploceras* **Buckman, 1913**
Order: Ammonitida
*Passaloteuthis* **Lissajous, 1915**
Order: Belemnitida
*Pleurotomaria* **Defrance, 1826**
Order: Murchisoniina
Appearance and lifespan: Middle Jurassic, 176 to 168 MYA. Location: Inferior Oolite, Burton Bradstock, Dorset, England, United Kingdom. Size: ammonite is 5 cm.

# What You End Up Collecting

I t's been said that interesting people collect interesting things. It can be ammonoids, other fossils, state pennies, or motorcars; it doesn't really matter what it is. If one is an object, and two is a pair, then three or more is a collection. Once you've herded together a few related items of any kind, you are compelled to learn about them, which likely results in you becoming a more fascinating and erudite human being. Perhaps collecting things that interest us is an essential ingredient in what makes us human. Conceivably, it appeals to our hunter-gatherer instinct.

"When I was seven years old, I was riding to school in a minivan that was used to transport kids in Manhattan," reminisced world-renowned trilobite authority Andy Secher. "I was sitting next to the driver, who happened to have an interesting rock on the car dashboard. I asked him what it was, and he showed me a Devonian-age coral that he had picked up during a visit to Albany the previous weekend. He gave me the rock, which he could tell fascinated me. That started me on my life-long collecting odyssey. Come to think of it, I still have that coral in my collection someplace!"

The fortunate ones are introduced to fossil collecting when they are still impressionable kids. You know how that works: a family member decides to drive a carload of cousins to a way-off-the-map location with the express intention of letting them play in the dirt with hammers and chisels. With a bit of luck, such an experience will result in them finding some cool fossilized stuff to take home. It is a truly rewarding

The search for fossils brings you to some extraordinary locations that you would otherwise have no reason to explore, like the slopes along Cottonwood Creek, Northern California.

*Speetoniceras Spath,* **1924** with pyrite
Order: Ammonitida. Appearance and lifespan: Early Cretaceous, 136.4 to 130.0 MYA. Location: Ulianorsk, Volga River, Russia. Size: 21 cm.

experience breaking apart a rock to discover a treasure that no human eye has ever seen before.

With such opportunities all around us, it is easy to understand why kids get hooked on paleontology. "My uncle Bob got me into collecting when I was eight years old," recalled fossil retailer Jeff Marshall. "We would collect all over Wisconsin. Every time we went looking, it was like going on a treasure hunt, and whenever you discovered something, it was as if you won a prize. We always had a great time, even if it was cold or muddy. It was such a party; I made a business out of it."

## Expertise

Fossil collecting is a relaxing, rewarding, and thoroughly enjoyable hobby. Ammonoids are particularly satisfying. There are about 20,000 species,

reflecting this cephalopod's nearly 400 million years on Earth, so there's always something new to locate—either online or in the field. Preservations can be found when you know where to look. The hunt for the physical specimen may take you to some extraordinary, off-the-beaten-track locations. Then again, if you are looking for something in particular, something exotic or magnificent, simply put some details in a search engine and relish the simplicity of the armchair experience. "When you have hands as soft as a baby's bottom, collecting fossils online is far more comfortable than actually venturing into the field. And by collecting in that manner, you're guaranteed not to get skunked by a fruitless day in the great outdoors," noted the urban-dwelling Secher.

Because ammonites were distributed worldwide throughout much of the Paleozoic and all of

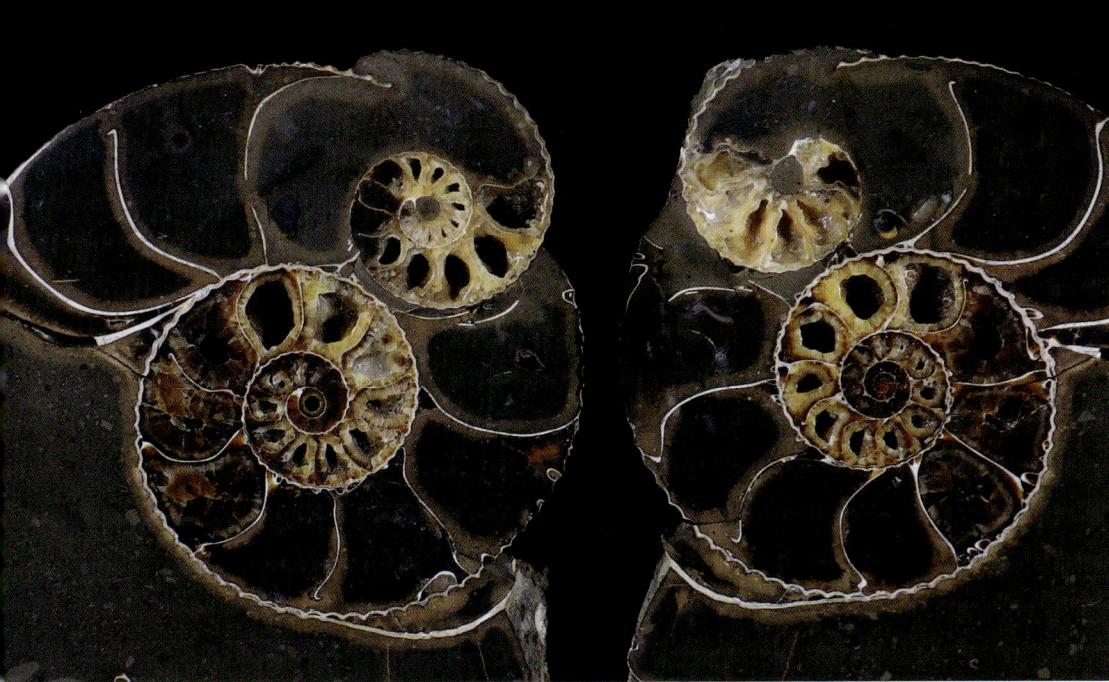

*Hoploscaphites* **Nowak, 1911**
Order: Ammonitida. Appearance and lifespan: Late Cretaceous, 70.6 to 66.0 MYA. Location: Elk Butte Member, Pierre Shale, Dewey County, South Dakota, United States. Size: 9 cm.

the Mesozoic eras, a cephalopod location tends not to be too far away (unless you live in a Cenozoic location or a glacial moraine). Get dig details through your local paleontology club, college science department, or research online—the fossil forums have valuable information. Map out your adventure. Use a sturdy backpack and fill it with the necessary tools: a rock hammer, chisel, and a pair of goggles to shield your eyes from flying fragments, and dental tools to excavate delicate areas. Then, for transporting fossils on the drive home, use aluminum foil for wrapping and packing containers for transportation. Bring a hard hat if you plan to venture under cliffs. Wear sturdy shoes and weather-appropriate outdoor gear. It's a great day for a picnic lunch, and, of course, bring plenty of precious water. Load up the car and away you go on a new adventure.

## Vocational Paleontology

When it comes to ammonoids, there are local experts who are rich with the secrets and stories of their region. By birth or by choice, professional and part-time paleontologists have become authorities by being based near extraordinary fossil locations. Kids from Somerset who kicked geodes along England's Jurassic Coast have become the most interesting tour leaders and museum docents. Curious schoolchildren who discovered fossils around the pinnacles of the Badlands of South Dakota in the United States have grown into world-renowned paleontologists. Youngsters who questioned the Devonian treasures emanating from Morocco's Tafilalt Plateau are some of today's most prolific commercial vendors, and kinder who played in the limestone canyons hiding Jurassic seas near Solnhofen, Germany, are now professors. Inquiring young minds around the world have evolved into the current crop of field authorities. "I started collecting minerals at age six. My parents took me to a mineral show," recalled Andreas Kerner, president of International Fossil Co., Inc. "The area I grew up in is rich in basalt (that is, agates, amethyst, et cetera), and I was able to hit some quarries with my dad on weekends. Found my first fossil at age nine, still have it. When I was twelve, I met some people who seriously dug in Permian sites, and tagged along as cheap labor."

Then there are those who come of age in less prolific locations. Perhaps they started with sea-shells and then expanded into Ammonoidea. The lack of field opportunities shifts them to research-ing, studying, better understanding the specimens at hand, and purchasing new fossils. Scholars specialize in specific areas of study. Interests like suture patterns or variance between ammonoid shapes will keep you in the lab or in the field look-ing for more species for comparison. Location proliferations and in situ phenomena offer field study opportunities. "Being a scientist is a won-derful experience. It gives you a chance to explore the world and learn so much about it," praised Austin Hendy, curator of invertebrate paleontol-ogy at the Natural History Museum of Los Ange-les County.

Ammonite collecting and research opportu-nities are available around the world, with many regions offering unique cephalopod-bearing for-mations. Each area captures a different time and place in paleontology history. In Pakistan or India, one would likely focus on the ceratites of the Salt Range. In upstate New York, someone leans toward goniatites. Around the Rocky Mountains and Canada, your backyard was once the West-ern Interior Seaway. North Americans along that inland ocean corridor often have a natural fascina-tion for Cretaceous cephalopods and likely dino-saurs, too. Some sophisticated collectors have a passion for wild, loosely coiled, and exotic heter-omorphs, while others investigate death pathol-ogies or soft-body bits. "I am passionate about healed pathologies as well as diagnosing bites," enthused Neal L. Larson, geologist and paleon-tologist at Larson Paleontology. "The fascinating thing about collecting and preparing Fox Hills and Pierre Shale ammonite concretions is that I never know what new discoveries we may find next. And those rocks contain some of the best

*Cleoniceras* ammonite cluster.

*Cleoniceras* **Parona & Bonarelli, 1897**
Order: Ammonitida. Appearance and lifespan: Late Cretaceous, 105 to 100 MYA. Location: Mahajanga, Madagascar. Size: largest is 12 cm.

and most beautifully preserved iridescent shell in the world."

Andreas Spatzenegger calls himself an advanced collector with two motives for selecting his ammonoid specialization. "The main reason for my passion in Triassic ammonoids is that the old Austrian 'Vienna School' paleontologists of the nineteenth century laid the cornerstone for our upper Triassic marine time scale based on ammonoids," he explained. "Since the second half of the nineteenth century, more than five hundred Triassic ammonoid species have been described from several Hallstatt limestone-successions in the Northern Calcareous Alps of Austria and Southern Germany: Hauer, 1848, 1855, 1860; Dittmar, 1866;

Hungary has put an ammonite on a national stamp.

**Reineckia crassicostata**
Face value: 2 forints. Country: Hungary. Issued: 1969.

The Hallstatt Limestone is one of the richest Triassic ammonite preservations on the planet. To date, it has yielded specimens from more than five hundred ammonoid species.

*Arietoceltites* **Diener, 1916** (sp)
*Anasirenites* **Mojsisovics, 1893** (sp)
*Megaphyllites* **Mojsisovics, 1879** (sp)
*Arcestes* **Suess, 1865** (sp)
*Cladiscites* **Mojsisovics, 1879** (sp)
Order: Ceratitida. Appearance and lifespan: Late Triassic, 225 to 216.5 MYA. Location: *Discotropites plinii* Ammonoid Zone, Hallstatt Formation, Upper Austria. Sizes: 2 to 3 cm.

Mojsisovics, 1873, 1889, 1893, 1895, 1896; Diener, 1921, 1926, et cetera. These monographic works and biostratigraphic postulations have since been transferred to other European regions such as the Appenines and Dinarides and, afterward, adopted and modified for North American and Asian sedimentary series. They build the fundamental cornerstones of today's marine Upper Triassic time scale, which is mainly based on ammonoids.

"The second reason is that these historic locations of the abovementioned authors are near to my home. So I can compare my finds with the first descriptions, search at historic locations that have been known for more than 150 years and track the trails of the old masters that lived in the former Austrian-Hungarian Empire."

## Avocational Paleontology

Many passionate rockhounds have the fantasy of discovering a fossil that changes our knowledge of the world. Think of the bragging rights one can claim from uncovering ammonoid soft-bodied preservations from Solnhofen, revealing the aggregation of *Discoscaphites* found above the iridium band in central New Jersey, or finding the juvenile feathered dinosaurs of the Gobi Desert in Mongolia. The prospect of discovering something awesome is open to everyone willing to look for it. There is limited funding for paleontological exploration, so many discoveries are made by avocational paleontologists and local field experts. All it takes is proper recording to confirm the identification. Patrick Hsieh, a board member of the Southern California Paleontological Society, has noted seven pieces of information that should be included to ensure the scientific importance of the fossils you find:

1. A description of where it was found, along with Global Positioning System (GPS) coordinates, if available.
2. Who collected it? If it's a new discovery, take the credit. The ammonoid genus *Meekoceras* is named for Indiana-born invertebrate paleontologist Fielding Bradford Meek; *Quenstedtoceras* honors Friedrich August Quenstedt, German mineralogist and paleontologist. Closer to home, *Baculites larsoni* is named in honor of Neal L. Larson.
3. When was it collected?
4. A way for someone else to identify your specimens. Photo references are good, as is a unique identification number for single specimens or for bulk specimen lots.
5. Describe what you have found to the best of your ability. Use descriptive words like "involute," "ribs," "nodules," and "spines."
6. Record the geological formation where you found the fossil, and add the layer if you know it, for example, Pumpkin Creek Formation, Daube Ranch near Ardmore, Oklahoma.
7. Add the age. Sources such as the United States Geological Survey (USGS), Mindat.org,

Fossilworks.org, and the Global Biodiversity Information Facility (gbif.org) are some of the sites that allow you to verify the stage, age, and period of your preservation.

"Absolutely, there are more great locations waiting to be found," enthused vocational paleontologist Kerner. "Many times, construction of a new housing development or freeway will expose new places. In the United States, the Grey Fossil Site (Washington County, Tennessee) is a good example. Road construction cut into an incredibly rich late Miocene deposit that, by all accounts, shouldn't have been there."

## Acquisitional Paleontology

Take notice that it is not uncommon to find a decorative ammonoid mounted on a pedestal on display in someone's home or office. Interesting cephalopod pieces can regularly be observed in the background on television shows and movie sets. You know that they didn't go to Madagascar or Morocco to find the fossils they have on display. If your time is valuable or you desire specimens that are prepped and likely more impressive than the ones you find, you can always buy your preservations. Many towns have fossils shops, which is where you can find familiar mementos such Moroccan ammonoids, Green River fish, Arkansas crystals, and perhaps exceptional local specialties. Or go online and look for fascinating finds. A general search for ammonites will lead you to the common iridescent Madagascan specimens that fill the sites. If you want something specific, try putting the genus name and / or location to uncover more unique materials. Opportunities are fleeting, but if you see a site that has sold the specimen you desire or has affiliations with the formation you're interested in, reach out to the vendor with your specific needs. Sometimes you win.

Writing on your finds is one way to identify your specimens. The Kanji writing adjacent to the HH8 on this Japanese Gaudryceras indicates the location it was found—Kamitombetsu, Hokkaido, while the symbols by the aperture of the preservation indicate that strata where is was found—below middle level.

**Gaudryceras de Grossouvre, 1894**
Order: Ammonitida. Appearance and lifespan: Late Cretaceous, 94.3 to 89.3 MYA. Location: Esashi, Hokkaido, Japan. Size: 7 cm.

If you hone your prepping skills, excellent uncleaned specimens can be had for less than the price of a six-pack of beer. Choose something in a soft matrix, and it can easily be worked into a display-quality fossil piece during your leisure time. "If someone wanted to give it a try on a smaller scale in front of the TV, they could use a magnifying visor and dental pick tools to slowly remove the softer matrix around the specimen," offered fossil preparator Ben Cooper. "The majority of the tools I use are pneumatic and can be quite noisy. I mostly listen to podcasts or books while working."

## ID Please

Once you start collecting things, the likely next step is figuring out what you have. Knowing a fossil's identification information guarantees that you can impress your friends and start a lively conversation. You will also understand where your preservations fall in the scheme of world evolution. The USGS and internet fossil forums are fine tools for this process. And if you are unsure, you can always add (sp.)—meaning that you know the genus but not the species. And if the genus is in question, include (?). With minimal time and effort, you will have acquired fascinating information about local paleontological esoterica. Keep interested in

Exotic ammonites and gastropod fossils for sale at the Tucson Gem and Mineral Show.

it long enough and you may turn into a specialist. You may find sharing your knowledge to be a satisfying experience.

## We Buy Ammonites: Tucson, Arizona

The Tucson Gem, Mineral and Fossil Show is equal parts nomadic souk, trade show, one of the world's best museums of fossils, and an international cache of natural resources. You can touch, fondle, and purchase everything from ammonoids to zeolite. Each year, for a three-week period beginning in late January, an international cadre of buyers and sellers—some 65,000 fossil aficionados —descend on this captivating, cactus-landscaped Southwestern outpost.

The fossil show is a unique experience; you will feel like you are Indiana Jones in search of treasures hiding in plain sight. As you ramble between hotel rooms and converted warehouses full of rock riches, you will rub elbows with paleontologists, diggers, jewelers, gemologists, researchers, museum curators, new age healy-feelies, fossil smugglers, crystal enthusiasts, dedicated hobbyists, casual shoppers, and bargain hunters, among others. Although everyone is interested in different items, each person has the same objective—to buy or sell unique items for the best possible price.

Once you realize the specific needs of your collection, take your shopping list and a set budget;

> **Ammonite Bite:** The term "keystone" may result in a wholesale discount. It is typically used for people who buy for retail resale. Depending on the scenario, you may need a wholesale tax license, but perhaps not. The discount may be as much as 50 percent off the marked cost. Do yourself a favor: when you are interested in a piece, ask, "Are the prices retail or keystone?"

otherwise, you will be so overwhelmed by all the incredible things you can actually own (from dinosaurs to *Didymoceras*) that you will run out of money long before you find what you intended to buy. There are boxes, flats, containers, and palettes from the four corners of the globe chock full of rocks, fossils, polished goods, salt lamps, and other unique specialties. You can find specimens from specific stages and locations as well as what is questionably the finest annual aggregation of ammonoids in the world. Once you've been to one Tucson show, you will want to return because each experience is so unique; there's always something fascinating and different. "It's the Superbowl of fossil shows," heralded Bill Barker of the Sahara Sea Collection and cofounder of the Mineral and Fossil Co-Op. In his space, you can find an amazing array of rare trilobites, ammonites, crinoids, and marine reptiles, and an impressive assortment of Moroccan polished goods. There's even a tabletop that delineates the evolution from the anoxic waters at the end of the Devonian to the more prolific waters of the Silurian Period.

The Tucson Convention Center hosts the main show, which is a gallery of gems and jewelry. Scores of satellite events, vendors, and pop-up retailers operate citywide. The fossil shows are traditionally on the west side of Tucson, near the I-10 freeway, but locations may vary. This grand celebration of Earth's treasures had an unassuming beginning back in 1955, when it debuted as a rock and mineral exhibition at a local elementary school. Today, it's the place where all that glitters is gold, pyrite, hematite, calcite, or some other exotic medium. If it is from Earth and it is of value, it's somewhere at the Tucson show, in an exhibition booth; a bulk bin; a flat under the presentation table; or a highly secured room, display case, or safe. The Tucson Gem, Mineral and Fossil Show offers something for all budgets and tastes—from teaching pieces of dinosaur bone to world-class

fossils, minerals, rocks, and preservations. It has become part of the city's economic fabric. The normally warm winter weather offers a respite from northern climes, making this desert fossil bazaar one of the highest revenue-producing events for the city's economy—annually generating upwards of $150 million.

## The Other Shows

The annual rock, mineral, and fossil circuit starts in early January, in Quartzite, Arizona. This massive stone flea market is very much an insider's event that everyone knows about. At this wholesale, pre-sale reunion caravansary, vendors will often buy, sell, and barter for merchandise they will unload at a premium several weeks later at the Tucson Gem, Mineral and Fossil Show. For many diggers from the snowy north, this desert oasis is a fine place to finish their prep and pad their showrooms before the big event. "It's a festival of rockhounds," summarized a participant.

As the year progresses, there are a series of three-to-seven-day fossil festivals in various international locations. At the start of summer, there is a big European fossil show in the charming French village of St. Marie aux Mines. Get there early because parking can be a challenge when this quaint town of five thousand people is overrun by the fossil and mineral elite. The school gym, the church courtyards, and the main streets are festooned with tents and tables sharing marvelous natural history pieces peddled predominantly

**Ammonite Bite:** Each culture has its own negotiation style. Some merchants haggle over price in any number of languages, while others may be insulted by such a request.

A box of acquisitions that boasts the spoils of several days spent at the annual Tucson fossil extravaganza.

by merchants from Europe, Africa, Asia, and the Middle East.

The East Coast Gem, Mineral and Fossil Show in Springfield, Massachusetts, is the locale for an August extravaganza. Early autumn brings the extensive Denver Mineral, Fossil, Gem and Jewelry Show, which occupies a convention hall and has satellite shows in historic buildings by the cattle stockyards. This event is followed by the Fossil Trade Fair hosted by the Mid-America Paleontology Society (MAPS) in Springfield, Illinois. (Did you know that there are currently 67 cities named Springfield in the United States?) A notable European fair to enjoy takes place in Munich, Germany, around the time of Oktoberfest. The Münchner Mineralientage is a well-organized fête featuring a vast array of hard-to-find local material. The enticing event draws naturalists from around the world to feed their fossil fantasies. Noteworthy gatherings also take place in Tokyo, Japan; Turin Italy; and almost every other developed country.

Numerous local and provincial bazaars are hosted by geology and paleontology groups and/or promoters. These are wonderful places to find specimens from the region of the show. Even though the internet has cut into everybody's bottom line by making material available to anyone, anywhere, at any time, these festivals are still a prime destination for the fossil elite looking for the exceptional pieces that never reach the internet or the mainstream. It is also a place to enjoy a unique sense of camaraderie among fellow enthusiasts and a chance to geek out on ammonoids.

## Auctions

Noteworthy pieces (fossils, gemstones, etc.) may be available at auction with tried-and-tested auction houses like Christie's, Sotheby's, Heritage, and Bonhams. These centuries-old firms comb the world for decorative material worthy of the elite. Ammonoids and dinosaurs marketed to art collectors are selling at prices far above their typical going rate. For example, "Apex," described as a coloring book Stegosaur, sold for $44.6 million at a Sotheby's in 2024. Stan, a forty-foot T-Rex sold for $31.8 million in 2020. Ammolite-clad Ammonoidea approach the $100,000 price point. These pieces go beyond the boundaries of paleontology when they cross over to the discerning eye of the art market.

Rare and extraordinary preservations, like this *Placenticeras* ammonite atop a bed of bivalves may reach their maximum potential sale price at an auction.

***Placenticeras costatum* (Herrick & Johnson, 1900)**
Order: Ammonitida. Appearance and lifespan: Late Cretaceous, 99.7 to 66.043 MYA. Location: undisclosed location, South Dakota. Size: ammonite, approximately 20 cm.

## The Darker Element

Get your fossils while you can because export laws are in constant flux. Countries like Canada, Argentina, Brazil, Kenya, South Africa, Mongolia, China, Australia, and various parts of Europe, among others, all have cultural heritage and national artifact laws. Although these laws largely pertain to antique objects, they are also designed to limit the export and sale of their natural resources (including vertebrate fossils, gems, and minerals). Ammonoids

are invertebrates and resemble seashells, so the rules don't tend to be quite as strict.

In the United States, rockhounds should be aware of legislation like the Omnibus Land Preservations Act, which, loosely interpreted, makes it illegal to buy, sell, or collect fossils on federal land without proper permitting. Several years ago, a smuggler was fined $25,000 for attempting to sell a 100-million-year-old *Psittacosaurus* (also known as "parrot lizard") fossil from Mongolia. The U.S. Department of Homeland Security regularly makes fossil-related arrests at the Tucson show. (Be wary of buying pterodactyl dinosaurs.) The website of the Association of Applied Paleontological Sciences is an excellent source for current laws and regulations regarding fossil ownership.

## China's Catch-22

When humans appeared in China approximately 1.7 million years ago, vertebrate bones were already part of the landscape. Having no idea what to make of these large bone shards sticking out of the hillsides, ancient people understandably misinterpreted them and assumed them to be the remains of dragons. As early as the fourth century BCE, a Chinese historian by the name of Chang Qu wrote about wyvern bones found in what we now call Sichuan Province. To the uneducated eye, a drum-sized piece of preserved whalebone or the scoot of a spiky *Stegosaurus* does indeed look like nothing else alive or formerly living on Earth.

Conflating dinosaurs and dragons is but one of the ironies of China's legacy with fossils. Another is that preservation formations are abundant in the People's Republic of China (PRC), but there are not enough qualified people to dig them up. Instead, fossils are excavated by locals seeking to sell their finds to earn a few extra yuan. This approach leaves many of the details desired by science in the dirt. "If it isn't collected right, a fossil loses its context—the

Sometimes preservations are exposed by wind and erosion. Buried fossilized animals form a slight mound and slowly weather out, while smaller bones are scattered around the location.

Image credit: Tom Horton from Shanghai, China, CC BY-SA 2.0, via Wikimedia Commons.

layer it was found in and its relationship to other fossils," observed Xu Xing, China's revered dinosaur pioneer from the Institute of Vertebrate Paleontology and Paleoanthropology in Beijing.

Toward the end of the twentieth century, the popularity of the *Jurassic Park* films drew hordes of young talent into professional paleontology throughout the Western world. Incongruously, that didn't happen in the PRC, where there are not enough students pursuing the bounty of the country's scientific riches. Even with the dearth of

paleontologists, China still lays claim to new species discoveries at an astonishing rate. It's a tough irony because the West acknowledges that the PRC has it all when it comes to dinosaurs. A Lagerstätte around Liaoning Province in the northeast has hundreds if not thousands of the Late Jurassic proto bird/dinosaur *Anchiornis.* In Hubei Province, the Cambrian Qingjiang biota soft-bodied preservations that date back 518 million years, older than Canada's acclaimed Burgess Shale. Other amazing fossil discoveries made in the PRC include some of the earliest animal embryos and deposits of mammals that tell the story of the rise of the Himalayas. In the southwestern part of the country, the Leye County of the Guangxi Zhuang Autonomous Region is acclaimed for its ammonoid biostratigraphy.

## We Visit Ammonites: China Has Layers

**Name: Shanggang Village,**
**Leye Fengshan UNESCO Global Geopark**
**Location: Leye County and Fengshan County,**
**Guangxi Zhuang Autonomous Region, Southwestern China**
**Age: Permian—Mesozoic**
**Date of UNESCO Geopark designation: 2015**

The Leye Fengshan UNESCO Global Geopark is notable for its awe-inspiring karst stone landscapes. These exceptional Paleozoic carbonate knolls are laden with ammonoids and other fascinating invertebrate fossils. Paradoxically, China promotes the site for the park's picturesque geology, dubbing it the "the territory of caves and of the world's longest natural bridge." Granted, the inspiring wind-carved mounds *are* an impressive site, A landscape of windswept hills, some up to 3,000 meters (9,843 feet) thick.

As the PRC has so many big, notable fossils, and there is a dearth of paleontologists, plants and

A panoramic view of Cuiping Village in Yangshuo, Guangxi, China.

Image credit: Chensiyuan, CC BY-SA 4.0, via Wikimedia Commons.

invertebrates are neglected. Prehistorians would so appreciate additional details about the Leye Formation's splendid ammonoid strata—which confirms the rapid evolution of cephalopod species in the early Triassic. Geologically, those were radical times. The devastating Permian–Triassic extinction was followed by the less lethal but truly relevant Smithian–Spathian boundary event some 251.9 million years ago. The catalyst was the eruption of a large igneous province in Siberia.

As a result of this molten activity, water temperature rose for the next 2 million years, with the surface temperature of tropical seas peaking at a spa-friendly 40°C (104°F). Among immobile fauna like molluscs and crinoids, only those that could cope with the heat survived. Half of all bivalve habitats disappeared. Fish remained and grew to

(right) *Paraceltites* **Gemmellaro, 1887**
Order: Ceratitida. Appearance and lifespan: Permian, 272.5 to 265.0 MYA. Location: Maokou Formation, Huangchang Ping Village, Jianshi County, Hubei, China. Size: largest is 6.5 cm.

The Guangxi region in southeastern China borders Vietnam.

Image credit: TUBS, CC BY-SA 3.0, via Wikimedia Commons.

**Ammonite Bite:** A slow-cooked ammonoid animal probably tasted like a brine-basted squid in a kelp broth.

impressive sizes as they trawled the seas for suffering prey. On land, large mobile vertebrate species migrated to cooler climates, while the steamy jungles became nearly devoid of animal life.

A versatile subclass until they weren't, ammonoids again proliferated after this stage's Early Triassic climactic devastation. They rapidly radiated to fill different oceanic niches around the now stabilized Tethys Sea. This evolution is there to inspect and calibrate at the Leye Formation. On the bottom tier of this exceptional repository, we find an attractive, frequently involute ammonoid with an arched keel. Known as *Meekoceras*, this simple, discus-shaped ceratite appeared about

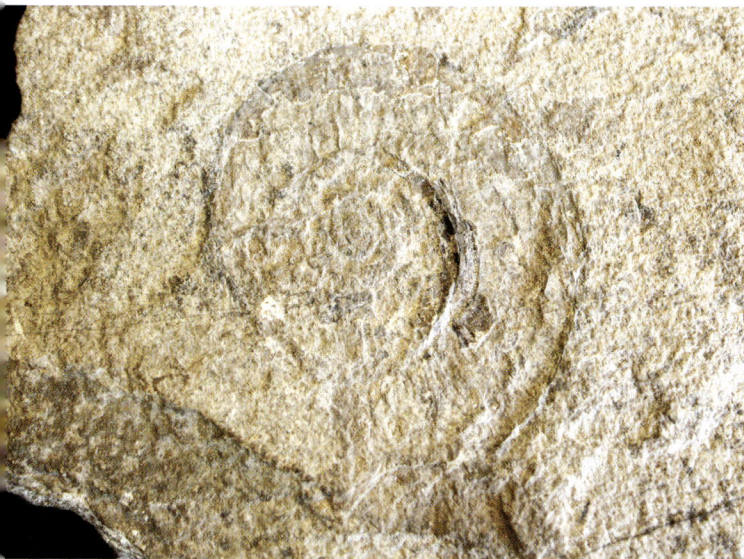

***Meekoceras gracilitatis* White, 1879**
Order: Ceratitida. Appearance and lifespan: Early Triassic, 251.3 to 247.2 MYA. Location: Paris Canyon, Meekoceras beds, Bear Lake County, Idaho, United States. Size: 4 cm.

*(right)* ***Xenodiscus agusticostatus* Welter, 1922**
Order: Ceratitida. Appearance and lifespan: Early Triassic, 251.3 to 247.2 MYA. Location: Tenggara Province, Indonesia. Size: largest is 2 cm.

252 million years ago and lasted no more than 5 million years. It was often accompanied by *Anasibirites*, which looked somewhat boxy because the keel was separated from the flanks by angular shoulders. Ribs tended to be unequal in length and depth, often with two fainter ridges grown between two stronger ones. The combination identifies the *Meekoceras* Zone. Biostratigraphers use this unit of cephalopod-bearing rock to correlate the age of similar occurrences. Beyond the Leye Formation in Guangxi, China, this seam has been identified in Siberia, Japan, Timor, New Zealand, the Himalayas, India, Pakistan, Madagascar, the northern Caucasus Mountains, and the Canadian Arctic, among other locales.

**Ammonite Bite:** A corresponding black band of the same stratigraphic age has been observed at the base of the Niti Limestone of the northern Indian margin in Spiti Valley, suggesting a short, Tethys Sea–wide, environmental event prior to the return to oxygenated bottom waters.

Atop the *Meekoceras* Zone is a narrow, ammonoid-rich bed from approximately 249 million years ago known as the *Flemingites* Limestone. The evolute ammonoids of this genus had flattish coils that gradually rose above the preceding whorl. Fine spiral ridges laced across the shell, while robust ribs flanked the sides. The ceratitic sutures were rounded like puzzle pieces. Above this horizon are the *Owenites* beds, a ribbon that has tiny calcite crystals and small ammonoids, indicating that cephalopods of this epoch suffered a premature demise followed by rapid fossilization. The narrow black shale band of mass death spreads across the entire Leye Platform and confirms the final pulse of the Siberian Traps some 247 million years ago.

At this formation in southwestern China, one can clearly spot the line where the eruptions end and marine oxygen again became plentiful. In the matrix are evolved forms of marine fauna and flora. This change is marked by cosmopolitan ammonoid genera that inhabited more temperate waters—Xenoceltitidae. A sophisticated spiral, it evolved from the earliest ceratites in the Late Permian Period. It's provenance, the superfamily Xenodiscoidea, provided the root stock for the subsequent expansion and diversification of ceratites in the Triassic. These evolute ammonoids were often found with the wavy-ribbed family member *Glyptophiceras*, and in some biotas there's *Bajarunia*, an offshoot of the *Meekoceras*. This layered cephalopod extravaganza was topped by *Tirolites* and *Columbites* and sprinkled with microfauna, which was preserved some 247 million years ago.

Seek and ye shall find and delight in the ammonoid strata of the Leye Fengshan Geopark. You may need to converse with several site custodians to get directions to the best cephalopod viewing spots, but don't be undaunted. When pondering these layers of history, one feels a powerful link to our planet, its past, and the ancient seas.

## Government Control

China's paleontological riches and their potential cannot be overestimated, but the country needs more qualified people to collect and identify the country's fossils. True to the PRC's entrepreneurial spirit, buried treasures have turned legions of farmers, construction workers, and even provincial bureaucrats into amateur fossil hunters angling to supplement their salary in the underground economy. But they are day traders in China's fossil ecosphere. The country deserves to have more professional paleontologists participating in the excitement of fossil excavation and research. Each season, elements disintegrate because no one collects them, even though there are willing volunteers. Researchers from institutions around the world wish to work on vertebrate preservations from the PRC, but the country's legislation makes all unauthorized digging for fossils an act of theft. Article 33 of the Cultural Relics Law explicitly forbids foreigners from prospecting or excavating within China's boundaries without special permission from the State Council. And the government generally grants very few authorizations to foreigners in any given year. Those who are caught exporting materials face severe sanctions. There are several urban legends of friends and exporters who have been jailed for shipping goods out of the PRC.

A simple way around these protocols is to buy a Chinese fossil online and have it sent to you. Yes, it may not be beneficial for the exporter, but an e-purchase is just another internet transaction.

*Trachyceras multituberculatum*
**Hsu, 1940**
Order: Ceratitida. Appearance and lifespan: Late Triassic, 235, to 222 MYA. Location: Lower Xiaowa Formation, Guanling, Anshun, Guizhou, China. Size: largest is 6 cm.

**Ammonite Bite:** The most effective way to check if a fossil is legally removed from the People's Republic of China is to look for the unique Ministry of Land and Resources registration number. This valuable set of digits confirms that the piece has officially left the country. If your Chinese fossil material does not have this number, then it is appropriate to be concerned as to just how it came to leave the PRC.

## The Legalities of Fossil Prospecting in China

Taking fossil finds away from China is a lot more difficult than fastening some bungee cords around your carton of preservations and telling the person at security check that you're a science geek bringing home a box of rocks. Under the PRC's laws, the first step for the transport of all material abroad is to report the items to the National Administration of Customs. From there, the responsibility of assigning export privileges shifts to the State Administration of Cultural Heritage. An appointee at this bureau forms a committee to examine and assess the material. (If you want to get fossils out of China, these are your new best friends.) Should your items qualify for export, the State Administration of Cultural Heritage issues a permit so that the specimens may proceed through the customs office to the designated port. No findings of scientific importance may go abroad without approval, and its common knowledge that the PRC issues permits to foreigners only under exceptional circumstances.

Looking like a slab of pizza with upside down mushrooms and stretchy cheese on it, this Permian ceratite plate is a unique and special fossil cephalopod preservation.

*Pernodoceras robustum* **Zhao & Liang, 1974**
Order: Ceratitida. Appearance and lifespan: Permian, 254 to 252 MYA. Location: Qingyuan, Guangdong, China. Size: largest is 5 cm.

## Still Moving Forward

In China's present social environment, it is normal for children to like dinosaurs, but adults prefer and revere dragons. The public's current interest in paleontology isn't big, and fossil experts are hard to come by. Where there are museums—Liaoning Province has about a dozen in various stages of development—there is often a lack of professionally trained employees to build education programs. The same consumerism that has catapulted Chinese buyers to the top of the luxury goods market makes lower-paying academic jobs and paleo history positions unappealing. Unlike more capitalistic professions such as economics, law, or computer science, the study of fossils doesn't benefit the development of modern Chinese society. A Sino survey reveals that paleontology ranks nearly 500th out of 1,433 college majors, just below theater director and above Sinhalese literature.

The breathtaking Himalayas.

Image credit: © Michel Royon, CC0, via Wikimedia Commons.

## We Dig Ammonites: Tibet

If you must excavate Chinese fossils or are looking for a side trip on your way to scale Mount Everest or visit Bhutan, Tibet may be an ideal stop for finding ammonoids. Rich in paleontology and history, and with a struggling economy, Tibet is a good choice should you know someone who wants to sidestep Sino bureaucracy and send fossils to themselves. The worst-case scenario is that the mailed items may not arrive at their destination, but no one will get in trouble for exporting them either. "Fossils from everywhere—even China— are available on outlets like eBay," noted acquisitor Andy Secher. "You just have to be prepared to wait three or four months for them to arrive in the mail. Though it may take your packages a long time to show up, 98 percent of the time, you will receive them."

Tibetan ammonoids are scattered in pockets amid the awe-inspiring, snow-covered peaks of the Himalayas. Located in a rugged, sparsely populated area with a short digging season, the Tulong Formation in southern Tibet preserves a well-defined ammonoid succession that's been in situ for approximately 248 million years. Tucked away in mountain crevasses and built into the sides of steep slopes are zones of ammonoid-rich mudstone that was once the bottom of the Tethys Sea.

**Sonninia (Euhoploceras)**
Order: Ammonitida. Appearance and lifespan: Middle Jurassic, 171.6 to 167.7 MYA. Location: Zhamunaqu Formation, Amdo County, Nagchu Prefecture, Tibet, China. Size: 6 cm.

The curious variety of Triassic ammonoid species discovered around the Tulong Formation represent different patterns of cephalopod distribution. Some cosmopolitan taxa from this sequence had broad geographic circulation. Search for the small involute ammonoid *Owenites*. *Paranannites* also hid its whorls but had a wider umbilicus. *Wasatchites* was evolute with a ribbed keel and a center that disappeared into the shell. Other Himalayan cephalopod beds are laden with more regional endemic species. You may find several genera from the family Arctoceratidae; it was a ceratite with a discus-shaped involute shell hiding compressed whorls that are taller rather than broad.

The Tulong ammonoid succession corresponds with Triassic cephalopod conglomerates from the other side of the Himalayas. Localized genera such as *Nammalites*, *Brayardites*, and *Truempyceras* have also been found in Pakistan's Salt Range, the Spiti Formation in India, the Oman Mountains, and South China. Earlier ammonoid faunas, like those found in the Leye Fengshan Geopark, are almost absent in Tulong because of the dearth of carbonate rocks.

If you made it this far, thank you for reading. It's been a pleasure and an accomplishment to publish this cephalopod compendium. I hope you enjoyed *We Dig Ammonites*.

# Closing Words

Ammonites allow us to touch immortality.

It is profound to hold the fossil of a creature

that was alive some 250 million years ago.

It bonds you metaphysically to the expanse

and depth of time in our seemingly

perpetual universe.

# Epilogue

Neal L. Larson

As someone who has been an ammonite fanatic for nearly my entire life, it is exciting to see something new and different published on the subject. There is something about ammonites that have fascinated and attracted people all around the world for centuries and will most likely continue for generations to come. Like so many, Jodi Summers became passionate about these beautiful, unique extinct creatures, which gave her the inspiration to take you along on a trip around the world as she embarked in her quest for ammonites and where they are found.

Humans are extremely curious beings, and ammonites are the perfect subject for study. Ammonites intrigue us, test us, enlighten us, confuse us, and baffle us. Those of us who work on ammonites spend our lives trying to understand them, but we never really do. How can we fully understand them when there is so much we don't know? How can they still be so mysterious when we have so many millions of specimens and have studied the rocks where they are found extensively? We know where and when they lived and identified the other animals that lived with them and ate them, but there is so much more to discover.

No other major taxa holds and hides so many secrets. Essential facts we should know but we don't include details like how similar were they to modern cephalopods? Were all ammonites carnivorous? Were they good swimmers? Did they have good vision? How rapidly did they grow? What was their life span, and how did they breed? Did they have chromatophores to enable them to change their color to alert for danger, express their fear, or show off?

Ammonites survived in the paleontological record for so long, but why did they go extinct when squid, octopuses, and nautiluses did not? How many arms (or tentacles) did ammonites have, and did their suture patterns have any relation to the number or shape of their arms? Out of the millions of ammonites collected worldwide, why haven't we yet found a single specimen with its arms, tentacles, eyes, hyponome, or any soft-body tissue preserved? Why have researchers discovered some of the inside stuff but none of the outside flesh? These and many, many more questions are still left unanswered.

You are never too young or too old to meet other ammonite-loving people. For some of us, it was a hobby that got "a little out of control." Going to the fossil shows introduced me to so many other devotees who share mine or similar passions. It's how I met Jodi and companion Andy Secher (a trilobite addict) years ago. It is an opportunity to make new friends and find new pieces for your collection from locales around the world. You will find that owning ammonites can bring you joy and possibly forever change your life.

This enlightening book will take you on a journey around the world, to ammonite localities that most will only dream of. It can open a wonderful hobby or career. I encourage everyone to take the time to read each and every page as you follow in Jodi's footsteps traveling to these unique and wonderful places in Earth's magnificent past, one ammonite after another. Maybe you, your child, or your grandchild will dig for ammonites and someday be able to unlock some of their hidden secrets.

NEAL L. LARSON
Larson Paleontology Unlimited
Geologist, paleontologist, collector, preparator, researcher, author, and confessed ammonite addict

# Acknowledgments and Thank Yous

Los Angeles has a unique set of seasons: summer, fire season, rainy season, and award season. If you live upon the Cenozoic sands that lap the Pacific Ocean, you are undoubtedly impacted by Hollywood. The *biz* spends the entire winter season rewarding the current crop of celebs for how fabulous they are and creating traffic snarls all around town. Let's go with this hoopla. For the thank yous and acknowledgements for *We Dig Ammonites*, imagine me, typically a lowkey science geek, standing onstage in some overdecorated yet unrecognizable theater. A team of stylists and makeup artists have made me look absolutely fabulous, and I am delighted as I accept an award for this cephalopod compendium from a jury of my peers. Turning on the charm, I thank the Academy of Paleontology and Earth Sciences (or whatever fictitious organization was doling out this particular award) and acknowledge the assistance of many colleagues and friends who had helped me reach this achievement. My appreciation is broadcast all over the world, including here, at the end of this book.

First and foremost, I would recognize Andy Secher, field associate at the American Museum of Natural History and author of several works, including *Travels with Trilobites* and *The Trilobite Collector's Guide*. Without his enthusiasm, support, and daily discussions, it would have been far more difficult to bring this project to fruition. A standing ovation for my longtime ammonite hero Neal Larson, geologist, paleontologist, and founder of Larson Paleontology; I was honored when he offered to edit this project. His efforts are greatly appreciated. Dr. Neil Landman, curator emeritus, fossil invertebrates, division of paleontology,

Richard Gilder Graduate School of the American Museum of Natural History in New York City, led me to many of the cephalopod-savvy natural history museum curators and university researchers cited in this book. He has made himself available for fruitful consultations and detailed explanations. Thank you to Miranda Martin, Kathryn Jorge, Milenda Lee, Robyn Massey, and the rest of the team Columbia University Press as well as Ben Kolstad of KnowledgeWorks Global for facilitating the realization of this project. I acknowledge my peer review committee for their valuable direction on early content. Dr. Christian Klug is to be commended for his valuable input on later drafts. To my family: their eyes don't always glaze over when I talk to them about ammonoids and other fossils. Media acknowledgments to Mindat.org, Fossilworks.org, and Wikipedia.org, among many other websites for supplying and verifying obscure information.

Now that the major players have been announced, let me offer kudos to those who contributed to this book. In recognition of their time and efforts, I would like to introduce you to this stellar array of professors, museum curators, email friends, researchers, collectors, prospectors, and commercial dealers, among others, and for their contributions to this cephalopod compendium. They are listed alphabetically, by first name:

Dr. Alexander Arkhipkin, senior fisheries scientist, Falkland Islands Fisheries Department

Andreas Kerner, president of International Fossil Co., Inc.

Andreas Spatzenegger, collector specializing in the history and ammonoids of the Tethyan Upper Triassic ammonoid zones

Andrzej Gorsky, engineer, collector, and field paleontologist

Dr. Anne Hildenbrand, biostratigraphy and palaeoecology research associate at the University of Heidelberg, Germany

Anthony Lindgren, founder of Lindgren Fossils LLC, second-generation commercial paleontologist specializing in Wyoming's Green River Formation

Aquarium of the Pacific

Dr. Austin Hendy, curator of invertebrate paleontology at the Natural History Museum of Los Angeles County

Ben Cooper, for his enthusiastic assistance in and explanations of fossil preparation

Bill Barker, president of Sahara Sea Collection, for always having a story and a goniatite to share

Bob Scriver, Blackfeet historian

British Geological Survey

Byron Blessed, geologist and paleobiologist for Natural Wonders Ltd.

Dr. Christian Klug, curator of the paleontological collections and exhibitions, University of Zurich, Switzerland

Christy Douglas, acquisitor

D. T. Donovan, Department of Geological Sciences, University College London

Daisuke Aiba, researcher at Fukada Geological Institute, previously of the Mikasa City Museum, Hokkaido, Japan

Dave Henderson, Lake Shasta prospector with a gift for spotting fossils, and trip photographer

David Temple, associate curator of paleontology, Houston Museum of Natural Science

David J. Peterman, Department of Geology and Geophysics, University of Utah

Dennis Veich, geology program manager emeritus for Shasta-Trinity National Forest

Dr. Dieter Korn, curator of the Museum für Naturkunde, Leibniz Institute for Evolutionary and Biodiversity Research, Berlin

Evelyn Siegfried, adjunct professor, Department of Archeology and Anthropology, University of Saskatchewan

Dr. Hans-Dieter Sues, senior research geologist and curator of vertebrate paleontology at the

Smithsonian's National Museum of Natural History, Washington, DC

Holly Walters, cultural anthropologist specializing in South Asian studies

Horace Parent, researcher, Paleontology Laboratory, Instituto de Fisiografía y Geología, Facultad de Ciencias Exactas, Ingeniería y AgrimensuraNational University of Rosario, Argentina

Japheth Boyce, Japheth Boyce Fossils and R.J.B. Rock Shop

Jason Cooper, prospector specializing in U.S. trilobites and dinosaurs. Thank you for the ammonite specimen.

Jeff Marshall, retail fossil dealer. Thank you for the many geologic treasures you have shared with me.

Joanna M. Wolfe, research fassociate, Javier Ortega-Hernández Lab, Harvard Museum of Comparative Zoology, Cambridge, Massachusetts, United States

John Catalani, member of the Mid-America Paleontology Society (MAPS)

John Cottrell, paleogeologist at the University of Rochester, New York, United States

John Issa, sales and marketing director at Enchanted Designs LTD, a Canadian company that specializes in ammolite ammonites

John Murray, tribal historic preservation officer for the Blackfeet Nation in Montana

José Juárez Ruiz, field paleontologist specializing in Western Europe

Joseph Grinnel, Wild West field biologist and zoologist

Joseph Huston, international heteromorph human

Dr. Joshua Slattery, Department of Physics, University of North Florida

Kate LoMedico Marriott, adjunct lecturer, Earth and Environmental Sciences, Brooklyn College, New York

Kathleen Ritterbush, associate professor of geology and geophysics at the University of Utah

Dr. Kenneth De Baets, research associate at the Institute of Evolutionary Biology, Faculty of Biology, University of Warsaw

Dr. Kimberly Meehan, chief scientist at Buffalo Museum of Science in Buffalo, New York, United States

Dr. Kirk Johnson, Sant Director of the Smithsonian National Museum of Natural History in Washington, DC

Lazare Eloundou Assomo, director of the UNESCO World Heritage Division

Dr. Lesley Cherns, honorary research fellow at the Cardiff University's School of Earth and Environmental Sciences, United Kingdom

Lizzie Hingley, founder of Stonebarrow Fossils

Mark Ault, scientific graphic designer

Mark Norell, curator emeritus, FARB, Division of Paleontology, and Professor Emeritus, Richard Gilder Graduate School, American Museum of Natural History, New York

Dr. Martin Shugar, trilobite collector

Dr. Melanie J. Hopkins, division chair, curator-in-charge, invertebrate paleontology, division of paleontology associate professor, Richard Gilder Graduate School, American Museum of Natural History, New York

Michał Zatoń, professor at the Institute of Earth Sciences, University of Silesia in Katowice, Poland

Mike Meacher, Stormbed Paleontological. Thank you for the Paleozoic ammonoids, penny photos, and shared wisdom.

Dr. Niles Eldredge, curator emeritus, division of paleontology, American Museum of Natural History, New York

Patrick Hsieh, vice president of the Southern California Paleontological Society

Dr. Paul D. Taylor, scientific associate and merit researcher emeritus at the Natural History Museum in London

Peter Harries, invertebrate paleontologist at the University of South Florida

Professor Randal Reed, chair of the Earth Sciences Department at Shasta College, California

Royal Mapes, professor of paleontology at Ohio University, and research associate at the American Museum of Natural History, New York

Dr. Rudolf Stockar, geosciences curator of the Cantonal Museum of Natural History in Lugano, Switzerland

Ryan Duffy, Lake Shasta prospector. Thank you for your unique specimen donation.

Sam Stubbs, esq., invertebrate paleontologist and attorney at large

Sania Arif, Department of Molecular Microbiology and Genetics University of Göttingen, Germany.

Steve Haire, Terrestrial Treasures

Troy Knowlton, First Nation Blackfoot tribal member

Xu Xing, deputy director, Institute of Vertebrate Paleontology and Paleoanthropology, Beijing

Thank you to the many science geeks and collectors everywhere who took interest in *We Dig Ammonites.*

# Bibliography/Additional Resources

## Chapter 1: The Aristocratic Ammonite

Andysfossils. "Yorkshire Ammonites (and Other Fossils) Revisited—A Knowledge Journey." Andysfossils.com, January 20, 2013. https://andysfossils.com/2013/01/20/lytoceras-or-visitors-from-the-deep/.

Bureau of Land Management, eds. "Can I Keep This? A Guide to Collecting on Public Lands." Bureau of Land Management, N.d. https://www.blm.gov/programs/natural-resources/forests-and-woodlands/forest-product-permits.

Butler, Patrick. "Fossils from Bacon Cove Could Rewrite History of Squid, Octopus." CBC/Radio-Canada, last updated January 26, 2022. https://www.cbc.ca/news/canada/newfoundland-labrador/fossils-nl-cephalopods-squid-octopus-1.6319018.

Chen, Y., C. Teichert. "Cambrian Cephalopods." *Geology* 11 (1983): 647–650. https://doi.org/10.1130/0091-7613(1983)11<647:CC>2.0.CO;2.

Cottrell, John. "Paleoecology of a Black Limestone Cherry Valley Limestone, Devonian, New York." University of Rochester, New York State Geological Association, 1972. https://ottohmuller.com/nysga2ge/Files/1972/NYSGA%201972%20G%20-%20Paleoecology%20Of%20A%20Black%20Limeston,%20Cherry%20Valley%20Limestone,%20Devonian,%20Ny%20State.pdf.

Di Cencio, Andrea, and Robert Weis. "Revision of Upper Toarcian Ammonites (Lytoceratidae, Graphoceratidae and Hammatoceratidae) from the Minette Ironstones, Southern Luxembourg." Travaux scientifiques du Musée national d'histoire naturelle Luxembourg, December 2020.

Eylott, Marie-Claire. "Mary Anning: The Unsung Hero of Fossil Discovery." The Trustees of the Natural History Museum, London. N.d.

https://www.nhm.ac.uk/discover/mary-anning-unsung-hero.html.

Gorsky, Andrzej. Engineer, collector, and field paleontologist. Email exchange with author, October 6, 2023.

Government of South Australia, eds. "Crown Land and Fossils." Department of Environment and Water, N.d. https://www.environment.sa.gov.au.

Gupta, Sanjeev; Jenny S. Collier, David Garcia-Moreno, et al. "Two-Stage Opening of the Dover Strait and the Origin of Island Britain." *Nature Communications* (2017). https://doi.org/10.1038/ncomms15101.

Harris, Paul. *Jurassic Coast: Britain's Heritage Coast.* Amberley, 2014.

Jurassic Coast World Heritage Site Partnership, eds. "What Is the Jurassic Coast?" Jurassic Coast World Heritage Site Partnership, 2022. https://jurassiccoast.org/.

Kaufman, Rachel. "Mary Anning: Life and Discoveries of the First Female Paleontologist." Future US Inc. and Live Science, February 22, 2021. https://www.livescience.com/who-was-mary-anning.html.

Love Lyme Regis Ltd., eds. "Fossil Hunting Activities." Love Lyme Regis Ltd., N.d. https://lovelymeregis.co.uk/fossils/fossil_hunting.

Lyme Regis Museum, ed. "Mary Anning's Story." Lyme Regis Museum, 20161. https://www.lymeregismuseum.co.uk/collection/mary-anning/.

McKeever, Amy. "What Are Ammonites, and How Did They Come to Rule the Prehistoric Seas?" National-Geographic.com, N.d. https://www.nationalgeographic.com/animals/facts/ammonites.

Meacher, Mike. Founding member, Stormbed Paleontological., Email dialogue with author, April 18, 2022.

Mindat.org. "Jurassic Coast World Heritage Site, England, UK." Hudson Institute of Mineralogy, January 15, 2023. https://www.mindat.org/loc-388536.html.

Newman, Cathy. "The Forgotten Fossil Hunter Who Transformed Britain's Jurassic Coast." National Geographic Society, March 29, 2021. https://www.nationalgeographic.com/travel/article/mary-anning-forgotten-fossil-hunter-british-jurassic-coast.

NOVA Science Trust, the Corporation for Public Broadcasting. "Ancient Builders of the Amazon." *PBS*, February 15, 2023. https://www.pbs.org/wgbh/nova/video/ancient-builders-of-the-amazon/.

Portail Orange. "Jurassic Ammonite Fossils from Central France." Portail Orange, N.d. http://jean-ours.filippi.pagesperso-orange.fr/anglais/domerienangl/lytoceratidaeangl.html.

Rafferty, John P. "Mary Anning." *Encyclopedia Britannica*, N.d. https://www.britannica.com/biography/Mary-Anning.

Sepkoski, Jack. "A Compendium of Fossil Marine Animal Genera (Cephalopoda Entry)." *Bulletins of American Paleontology* 363 (2002): 1–560.

Tappenden, Roz. "Ammonite: Who Was the Real Mary Anning?" *BBC*, October 17, 2020. https://www.bbc.com/news/uk-england-dorset-54510746.

Taylor, Christopher. "Ancyloceratidae." Variety of Life, December 10, 2016. http://taxondiversity.fieldofscience.com/2016/12/ancyloceratidae.html.

TheSchoolRun. "Mary Anning." *TheSchoolRun*. 2022. https://www.theschoolrun.com/homework-help/mary-anning.

Thompson, LuAnne. "Tide Dynamics—Dynamic Theory of Tides." University of Washington, March 3, 2016. http://faculty.washington.edu/luanne/pages/ocean420/notes/tidedynamics.pdf.

Trench, Tommy. "Salvaging Fossils on the Jurassic Coast." *Hakai Magazine*, May 26, 2020. https://www.hakaimagazine.com/videos-visuals/salvaging-fossils-on-the-jurassic-coast/.

Thecosmilia Trichitoma. "Fossil Laws in Germany." *Fossil Forum*, September 22, 2018. http://www.thefossilforum.com/topic/88581-fossil-laws-in-germany/.

UNESCO. "Dorset and East Devon Coast." *UNESCO. org*, n.d. https://whc.unesco.org/en/list/1029/.

Whicher, John T., Robert B. Chandler, and René Hoffmann. "Lytoceratid Ammonites from the Inferior Oolite Formation (Middle Jurassic, Aalenian and Bajocian) of Dorset (United Kingdom)." *Proceedings of the Geologists' Association* 134, no. 2 (2023): 216–245. https://doi.org/10.1016/j.pgeola.2023.03.001.

Zielinski, Sarah. "Mary Anning, an Amazing Fossil Hunter." *Smithsonian Magazine*, January 5, 2010. https://www.smithsonianmag.com/science-nature/mary-anning-an-amazing-fossil-hunter-60691902/.

## Chapter 2: Orders and Disorder

Arkell, W. J., W. M. Furnish, Bernhard Kummel, et al. "Treatise on Invertebrate Paleontology, Part L, Mollusca 4." Geological Society of America and the University of Kansas Press, 1957.

Arkhipkin, Alexander I. "Getting Hooked: The Role of a U-Shaped Body Chamber in the Shell of Adult Heteromorph Ammonites." Falkland Islands Fisheries Department, February 2014. https://www.researchgate.net/requests/r117247439.

Chapman, A. D. *Numbers of Living Species in Australia and the World*, 2nd ed. Australian Biological Resources Study, 2009.

Di Cencio, Andrea, and Robert Weis. "Revision of Upper Toarcian Ammonites (Lytoceratidae, Graphoceratidae and Hammatoceratidae) from the Minette Ironstones, Southern Luxembourg." *Travaux scientifiques du Musée national d'histoire naturelle Luxembourg*, December 2020.

Donovan, D. T. "History of Classification of Mesozoic Ammonites." *Journal of the Geological Society* (1994). https://doi.org/10.1144/gsjgs.151.6.1035.

Fedosov, Alexander E., and Nicolas Puillandre. "Phylogeny and Taxonomy of the Kermia–Pseudodaphnella (Mollusca: Gastropoda: Raphitomidae) Genus Complex: A Remarkable Radiation via Diversification of Larval Development." *Systematics and Biodiversity* 10, no. 4 (2012): 447–477. https://doi.org/10.1080/14772000.2012.753137.

Flower, Rousseau H. "The Nautiloid Order Ellesmerocerida." New Mexico Bureau of Mines and Mineral Resources. Memoir 12, 1964.

Flower, Rousseau H. "Saltations in Nautiloid Coiling." *Evolution* 9, no. 3 (1955): 244–260. https://doi.org/10.2307/2405647.

Flower, Rousseau H., and Bernhard Kummel Jr. "A Classification of the Nautiloidea." *Journal of Paleontology* 24, no. 5 (September 1950): 604–616. https://www.jstor.org/stable/1299547.

Furnish, W. M., and Brian F. Glenister. "Ellesmerocerida." In *Treatise on Invertebrate Paleontology, Part-K, Nautiloidea*, ed. Curt Teichert, et al. Geological Society of America and University of Kansas Press, 1964, K160–K188.

Government of South Australia. "Crown Land and Fossils." Department of Environment and Water, N.d. https://www.environment.sa.gov.au.

Hoffmann, René, Michael K. Howarth, Dirk Fuchs, Christian Klug, and Dieter Korn. "The Higher Taxonomic Nomenclature of Devonian to Cretaceous Ammonoids and Jurassic to Cretaceous Ammonites Including Their Authorship and Publication." *Neues Jahrbuch für Geologie und Paläontologie* 305, no. 2 (2022): 187–197. https://doi.org/10.5167/uzh-220683.

Klug, Christian, Dieter Korn, Neil H. Landman, Kazushige Tanabe, Kenneth De Baets, and Carole Naglik. "Describing Ammonoid Conchs." *Ammonoid Paleobiology: From Anatomy to Ecology* (August 2015). https://doi.org/10.1007/978-94-017-9630-9_1.

Klug, Christian, Dieter Korn, Kenneth De Baets, Isabelle Kruta, and Royal H. Mapes. "Evolutionary Patterns of Ammonoids: Phenotypic Trends, Convergence, and Parallel Evolution." *Ammonoid Paleobiology:*

*From Anatomy to Ecology* (August 2015). https://doi.org/10.1007/978-94-017-9633-0_5.

Klug, Christian, Björn Kröger, Jakob Vinther, and Dirk Fuchs. "Ancestry, Origin, and Early Evolution of Ammonoids." *Topics in Geobiology* 44 (August 2015). https://doi.org/10.1007/978-94-017-9633-0_1.

Kröger, Björn. "Some Lesser Known Features of the Ancient Cephalopod Order Ellesmerocerida (nautiloidea, cephalopoda)." *Paleontology* 50, no. 3 (2007); 556–572. https://doi.org/10.1111/j.1475-4983.2007.00644.x.

Kröger, Björn, et al. "Pulsed Cephalopod Diversification during the Ordovician." *Palaeogeography, Palaeoclimatology, Palaeoecology* 273, nos. 1–2 (2009): 174–183. https://doi.org/10.1016/j.palaeo.2008.12.015.

Landman, Neil H., Richard Arnold Davis, and Royal H. Mapes, eds. *Cephalopods Present and Past: New Insights and Fresh Perspectives*. Springer, 2007.

Mindat.org. "Phylloceras." Hudson Institute of Mineralogy, October 2, 2023. https://www.mindat.org/taxon-4622352.html.

Mutvei, Harry. "Characterization of Nautiloid Orders Ellesmerocerida, Oncocerida, Tarphycerida, Discosorida and Ascocerida: New Superorder Multiceratoidea." *GFF* 135, no. 2 (June 2013). https://doi.org/10.1080/11035897.2013.801034.

Peterman, David J., Kathleen A. Ritterbush, Charles N. Ciampaglio, et al. "Buoyancy Control in Ammonoid Cephalopods Refined by Complex Internal Shell Architecture." *Scientific Reports* 11, no. 8055 (2021). https://doi.org/10.1038/s41598-021-87379-5.

Ponder, W. F., and David R. Lindberg, eds. *Phylogeny and Evolution of the Mollusca*. University of California Press, 2008.

Prehistoric Planet Store. "Lytoceras Fimbriatum, Jurassic Ammonite." PaleoClones, N.d. https://www.prehistoricstore.com/item.php?item=3193.

Rosenberg, Gary. "Mollusckque—Mollusk vs. Mollusc." *Conchologists of America*, 1996. https://conchologistsofamerica.org/mollusckque-mollusk-vs-mollusc/.

Rosenberg, Gary. "A New Critical Estimate of Named Species-Level Diversity of the Recent Mollusca." *American Malacological Bulletin* 32, no. 2 (2014): 308–322. https://doi.org/10.4003/006.032.0204.

Sepkoski, Jack. "A Compendium of Fossil Marine Animal Genera (Cephalopoda Entry)." *Bulletins of American Paleontology* 363 (2002): 1–560.

Taylor, Christopher. "Ancyloceratidae." Variety of Life, December 10, 2016. http://taxondiversity.fieldofscience.com/2016/12/ancyloceratidae.html.

Wade, M. "Nautiloids and Their Descendants: Cephalopod Classification in 1986." Memoir 44, New Mexico Bureau of Mines and Mineral Resources, 1988.

Wiedmann, J. "The Heteromorphs and Ammonoid Extinction." *Biological Reviews* 44 (1960): 563–602.

Wright, C. W., J. H. Callomon, and M. K. Howarth. *Treatise on Invertebrate Paleontology pt L. Mollusca 4, Revised vol. 4. Cretaceous Ammonoidea*. Geological Society of America and the University of Kansas, 1996.

## Chapter 3: Named for a God

Algeo, T. J. "Terrestrial-Marine Teleconnections in the Devonian: Links Between the Evolution of Land Plants, Weathering Processes, and Marine Anoxic Events." Philosophical Transactions of the Royal Society B: Biological Sciences, 1998. https://doi.org/10.1098/rstb.1998.0195.

Allen, Gabe. "Marine Life Perished from the Late Devonian Mass Extinction, but Its Origin Is Still Shrouded in Mystery." Kalmbach Media Co., September 12, 2022. https://www.discovermagazine.com/the-sciences/the-late-devonian-mass-extinction-explained.

Amos, Jonathan. "Fossil tracks record 'oldest land-walkers.'" *BBC News*, January 6, 2010. http://news.bbc.co.uk/2/hi/science/nature/8443879.stm.

Baez, John. "Extinction." University of California Riverside, April 8, 2006. https://math.ucr.edu/home/baez/extinction/.

Barker, Bill. President of Sahara Sea Collection. Conversation with author, October 9, 2023.

Ben Yahia, Jihane. "Morocco's Surging Trade in Fossils." ENACT Africa, June 7, 2019. https://enactafrica.org/research/trend-reports/moroccos-surging-trade-in-fossils.

Carlson, Marvin P. "Transcontinental Arch—A Pattern Formed by Rejuvenation of Local Features across Central North America." *Tectonophysics* 305, no. 1–3 (May 1999): 225–233. https://doi.org/10.1016/S0040-1951(99)00005-0.

Catalani, John A. "A Brief Introduction to the Devonian Period." *Official Publication of the Mid-America Paleontology Society* 36, no. 2. (April 2013): 000–000.

Chlupáč, Ivo, and Jindrich Hladil. "The Global Stratotype Section and Point of the Silurian-Devonian Boundary." *CFS Courier Forschungsinstitut Senckenberg* 00, no. 1–8 (January 2000): 000–000.

Danilchik, Walter. and Shah, S.M. Ibrahim. "Stratigraphy and Coal Resources of the Makarwal Area, Trans-Indus Mountains, Mianwali District, Pakistan." U.S. Geological Survey, 79–532 (1987): 10.3133/ofr79532 https://doi.org/10.3133/ofr79532.

Dumper, Michael R. T., and Bruce E. Stanley, eds. "Cities of the Middle East and North Africa: A Historical Encyclopedia." ABC-CLIO, 2007.

Gradstein, Felix M., James G. Ogg, and Alan G. Smith. *A Geologic Time Scale 2004*. Cambridge University Press, 2004.

Gutiérrez-Marco, Juan Carlos, and Diego C. García-Bellido. "The International Fossil Trade from the Paleozoic of the Anti-Atlas, Morocco." *Geological Society*, Special Publications 485 (Month, 0000): 69–96. https://doi.org/10.1144/SP485.1, https://www.cambridge.org/core/journals/international-journal-of-astrobiology/article/abs/impact-of-tectonicstyle-on-marine-transgression-and-evolution/6FEEA5A54831EE672DE0982FA0B35318#.

Haq, Bilal U., and Stephen R Schutter. "A Chronology of Paleozoic Sea-Level Changes." *Science* 322, no. 5898 (October 3, 2008): 64–68. https://doi.org/10.1126/science.1161648.

House, Michael R. "Devonian Extinctions." *Encyclopedia Britannica*, March 26, 2009. https://www.britannica.com/science/Devonian-extinctions.

Iqbal, Shahid, Irfan U. Jan, and Muhammad Hanif. *The Mianwali and Tredian Formations: An Example of the Triassic Progradational Deltaic System in the Low-Latitude Western Salt Range, Pakistan*. King Fahd University of Petroleum and Minerals. 2013.

Kaim, A., A. Nützel, M. Hautmann, and H. Bucher. "Early Triassic Gastropods from Salt Range, Pakistan." *Bulletin of Geosciences* 88, no. 3 (2013): 505–516 (seven figures).

Kaiser, Sandra. "The Devonian/Carboniferous Boundary Stratotype Section (La Serre, France) Revisited." *Newsletters on Stratigraphy* 43, no. 2 (April 1, 2009): 195–205. https://doi.org/10.1127/0078-0421/2009/0043-0195.

Kaufmann, B., E. Trapp, and K. Mezger. "The Numerical Age of the Upper Frasnian (Upper Devonian) Kellwasser Horizons: A New U-Pb Zircon Date from Steinbruch Schmidt (Kellerwald, Germany)." *Journal of Geology* 112, no. 4 (2004): 495–501. https://doi.org/10.1086/421077.

Klug, Christian. Email exchange with author, August 10, 2022.

Korn, Dieter. "Ammonoid Evolution in Late Famennian and Early Tournasian." *Ministry of Economic Affairs, Administration of Mines, Belgian Geological Survey* 109, no. 1 (1986): 49–54.

https://www.researchgate.net/publication/259475024_Ammonoid_evolution_in_Late_Famennian_and_Early_Tournaisian/comments.

Leonova T. B. "Permian Ammonoids: Classification and Phylogeny." *Paleontological Journal* 36 (January 2002): S1–S114. https://www.researchgate.net/publication/287517715_Permian_ammonoids_Classification_and_phylogeny.

Neuendorf, Henri. "Illegal Trading Degrades World's Largest Fossil Museum." Artnet News, October 28, 2014. https://news.artnet.com/art-world/illegal-trading-degrades-worlds-largest-fossil-museum-146610.

Newitz, Annalee. "How Do You Have a Mass Extinction without an Increase in Extinctions?" *The Atlantic*, June 13, 2013. https://www.theatlantic.com/technology/archive/2013/06/how-do-you-have-a-mass-extinction-without-an-increase-in-extinctions/276836/.

Osborne, Lawrence. "The Fossil Frenzy." *New York Times*, October 29,2009. https://www.nytimes.com/2000/10/29/magazine/the-fossil-frenzy.html.

Paproth, Eva, Raimund Feist, and Gerd Flajs. "Decision on the Devonian-Carboniferous Boundary Stratotype." *Episodes* 14, no. 4 (December 1991): 331–336. https://doi.org/10.18814/epiiugs/1991/v14i4/004.

Parry, S. F., S. R. Noble, Q. G. Crowley, and C. H. Wellman. "A High-Precision U–Pb Age Constraint on the Rhynie Chert Konservat-Lagerstätte: Time Scale and Other Implications." *Journal of the Geological Society* 4 (2011): 863–872. https://doi.org/10.1144/0016-76492010-043.

Payne, J. L., J. A. Al Aswad, C. Deutsch, P. M. Monarrez, J. L. Penn, and P. Singh. "Selectivity of Mass Extinctions: Patterns, Processes, and Future Directions." *Cambridge Prisms: Extinction* 1, no. e12 (2023). https://doi.org/10.1017/ext.2023.10.

PrideofPakistan.com, eds. "Khewra Salt Mine Is Pakistan's Largest and the World's 2nd Largest Salt Mine." prideofpakistan.com, n.d. https://www.prideofpakistan.com/explore-pakistan-detail/Khewra-Salt-Mine-is-Pakistans-largest-and-the-worlds-2nd-largest-Salt-Mine/89.

Qureshi, Kaleem Akhtar, and Ali Nasir Fatmi. "Sedimentary Geology and Biostratigraphy of the Mianwali Formation, Kala Chitta Range, Northern Pakistan." *Nepal Geological Society* 19 (1999): 49–57. https://ngs.org.np/sedimentary-geology-and-biostratigraphy-of-the-mianwali-formation-kala-chitta-range-northern-pakistan/#comment-126.

Ramsar Sites Information Service, eds. "Oasis du Tafilalet." Ramsar Sites Information Service, January 15, 2005. https://rsis.ramsar.org/ris/1483.

Saneeni, S. J. "The Salt Range: Pakistan's Unique Field Museum of Geology and Paleontology." Notebooks on Geology, Brest, Book, March 2009. Chapter 6 (CG2009_BOOK_03/06).

Sicree, Andrew A. "Saudi Aramco World: Morocco's Trilobite Economy" Aramcoworld.com, 2009. Retrieved July 6, 2021.

UNESCO, eds. "The Salt Range and Khewra Salt Mine." UNESCO World Heritage Centre, n.d. https://whc.unesco.org/en/tentativelists/6118/.

Wicander, Reed, and James S. Monroe. *Historical Geology: Evolution of Earth and Life through Time*, 7th ed. Cengage Learning, 2013.

Zakharov, Yuri D., and Nasrin Moussavi Abnavi. "The Ammonoid Recovery after the End-Permian Mass Extinction: Evidence from the Iran-Transcaucasia Area, Siberia, Primorye, and Kazakhstan." *Acta Palaeontologica Polonica* 58, no. 1 (March 1, 2013):127–114. https://doi.org/10.4202/app.2011.0054.

## Chapter 4: Death and Fossilizaiton

American Geosciences Institute, eds. "How Do Paleontologists Identify Fossils?" American Geosciences Institute, n.d. https://www.americangeosciences

.org/education/k5geosource/content/fossils/how-do-paleontologists-identify-fossils.

Arthur, Damon. "Scientists Planning to Search for Triassic Fossils around Lake Shasta." Redding Record Searchlight, March 5, 2016. https://www.redding.com/story/news/local/2016/05/03/scientists-planning-to-search-for-triassic-fossils-around-lake-shasta/93714916/.

Briles, Christy E., Cathy Whitlock, Carl N. Skinner, and Jerry Mohr. "Holocene Forest Development and Maintenance on Different Substrates in the Klamath Mountains, Northern California, USA." *Ecology. Ecological Society of America* 92, no. 3 (2011): 590–601. https://doi.org/10.1890/09-1772.1.

Bureau of Land Management, eds. "Can I Keep This? A Guide to Collecting on Public Lands." Bureau of Land Management, n.d. https://www.blm.gov/programs/natural-resources/forests-and-woodlands/forest-product-permits.

Klug, Christian. Paleobiology department member, University of Zürich, Switzerland, Palaeontological Institute and Museum. Email exchange with author, March 6, 2023.

Coyne, J. A. "Evolution: Ernst Mayr (1904–2005)." *Science* 307, no. 5713 (2005): 1212–1213. https://doi.org/10.1126/science.1110718.

Editors of Encyclopedia Britannica, The. "Ammonoid." *Encyclopedia Britannica*, May 27, 1999. https://www.britannica.com/animal/ammonoid.

FossilEra.com, eds. "About Ammonites." FossilEra, 2023. https://www.fossilera.com/pages/ammonites.

Hyatt, Alpheus, and J. P. Smith. "The Triassic Cephalopod Genera of America." U.S. Government Printing Office, 1905. https://doi.org/10.3133/pp40.

Jurassic Coast Trust, eds. "Welcome to the Jurassic Coast." Jurassic Coast Trust Trading Co., 2023. https://jurassiccoast.org/fossilfinder/focus-on-fossils/molluscs/.

Lewon, Dennis. *Hiking California's Trinity Alps Wilderness*. Falcon, 2001.

Logan, Alan. "Triassic Period Geochronology." *Encyclopedia Britannica*, July 26, 1999. https://www.britannica.com/science/Triassic-Period#ref755343.

Milner, Richard. "The Encyclopedia of Evolution." Facts on File, 1990.

Noble, Paula, and Renne Paul. "Paleoenvironmental and Biostratigraphic Significance of Siliceous Microfossils of the Permo-Triassic Redding Section, Eastern Klamath Mountains, California." *Marine Micropaleontology* 15, no. 3–4 (1927): 379–391. https://doi.org/10.1016/0377-8398(90)90021-D.

Norman, D. B. "Mary Anning and Her Times: The Discovery of British Palaeontology (1820–1850)." *Trends in Ecology & Evolution* 14, no. 11 (November 1999): 420–421, https://doi.org/10.1016/S0169-5347(99)01700-0.

Proffitt, Pamela. *Notable Women Scientists*. Gale Group, 1999.

Sandoval, José, Luis O'Dogherty, and Jean Guex. "Evolutionary Rates of Jurassic Ammonites in Relation to Sea-Level Fluctuations." *Palaios* 16, no. 4 (2001): 311–335. https://doi.org/10.2307/3515574.

Sawyer, John O. *Northwest California: A Natural History*. University of California Press, 2006.

Skinner, Carl N., Alan H. Taylor, and James K. Agee. "Klamath Mountains Bioregion." Fire in California's Ecosystems. University of California Press, 2006. https://doi.org/10.1525/california/9780520246058.003.0009.

Smith, James Perrin. *Upper Triassic Marine Invertebrate Faunas of North America*. U.S. Government Printing Office, 1927.

USGS, eds. "Index Fossils." United States Geological Survey Publications Services, July 28, 1997. https://pubs.usgs.gov/gip/geotime/fossils.html.

Vendetti, Jann. "The Cephalopoda." University of California Museum of Paleontology, 2006. https://ucmp.berkeley.edu/taxa/inverts/mollusca/cephalopoda.php.

White, Mike. *Trinity Alps and Vicinity: Including Whiskeytown, Russian Wilderness, and Castle Crags Areas*, 5th ed. Wilderness Press, 2010.

Wuerthner, George. *California's Wilderness Areas*. Westcliffe, 1997.

## Chapter 5: Death and Fossilization

20minutos.es, eds. "La Unesco declara como reserva de la biosfera al bosque seco del sur de Ecuador." 20 Minutos (January 22, 2015).https://www .20minutos.es/noticia/2354160/0/reserva-biosfera /bosque-seco-ecuador/conservacion-ambiental/.

Andy. "Posidonia Shale Transcript and References." Fossilbonanza.com, October 28, 2020. https:// fossilbonanza.com/tag/holzmaden/.

Arkell, W. J., R. C. Moore, A. K. Miller, W. M. Furnish, and O. H. Schindewolf. *Mesozoic Ammonoidea. Treatise on Invertebrate Paleontology, Part L, Mollusca 4.* Geological Society of American and University of Kansas Press, 1957.

Banco Central del Ecuador, eds. "Sector Minero— Cartilla Informativa." Banco Central del Ecuador, February 13, 2015. https://contenido.bce.fin.ec.

BioAcid.com, eds. "Biological Impacts of Ocean Acidification." BioAcid, n.d. https://www.ocean acidification.de/calcite-aragonite/?lang=en.

Borenstein, Seth. "Hints of Life on What Was Thought to Be Desolate Early Earth." Associated Press, October 19, 2015. https://apnews.com/hints-of -life-on-what-was-thought-to-be-desolate-early -earth-e6be2537b4cd46ffb9c0585bae2b2e51.

Brenner, K., and A. Seilacher. "New Aspects about the Origin of the Toarcian Posidonia Shales." *Abhandlungen Neues Jahrbuch fur Geologie und Palaontologie*, 157, no. 3 (1978): 000–000.

Bulot, Luc Georges. Emmanuel Robert. Etienne Jaillard, and W. J. Kennedy. "The Albian Ammonite Successions of the Subandean Zone of Ecuador." Fourth European Meeting on the Palaeontology and Stratigraphy of Latin America Cuadernos del Museo Geominero, number 8, Instituto Geológico y Minero de España, Madrid (2007). https:// www.academia.edu/19519727/The_Albian _ammonites_successions_of_the_Subandean _Zone_of_Ecuador.

Clements, Thomas, and Sarah Gabbott. "Exceptional Preservation of Fossil Soft Tissues." eLS, April 13, 2022. https://doi.org/10.1002/9780470015902 .a0029468.

Daeschler, Edward B., Neil H. Shubin, and Farish A. Jenkins Jr. "A Devonian Tetrapod-Like Fish and the Evolution of the Tetrapod Body Plan." *Nature*, April 6, 2006. https://doi.org/10.1038 /nature04639.

Ebli, Oskar, István Vető, Harald Lobitzer, Csanád Sajgó, Attila Demény, and Magdolna Hetényi. "Primary Productivity and Early Diagenesis in the Toarcian Tethys on the Example of the Mn-Rich Black Shales of the Sachrang Formation, Northern Calcareous Alps." *Organic Geochemistry* 29, no. 5–7 (1998). https://doi.org/10.1016/s0146 -6380(98)00069-2.

Ecuador.com, eds. "Puyanop Petrified Forest, Protected Areas, Tourist Attractions." Ecuador. com, n.d. https://www.ecuador.com/attractions /protected-areas/puyango-petrified-forest/.

Fossiland.com, eds. "About Holzmaden." Holzmaden. com, 2023. http://www.holzmaden.com/.

Hoesch, William A. "Lagerstätten!" ICR Institute for Creation Research, August 1, 2007. https://www .icr.org/article/lagerstatten.

Instituto de Investigación Geológico y Energético Gobierno del la República del Ecuador. "Bosque Petrificado de Puyango camino a ser el segundo Geoparque Mundial del Ecuador." Instituto de Investigación Geológico y Energético, n.d. https:// www.geoenergia.gob.ec/bosque-petrificado -de-puyango-camino-a-ser-el-segundo-geoparque -mundial-del-ecuador/.

Jablonski, David, Kaustuv Roy, James W. Valentine, Rebecca M. Price, and Philip S. Anderson. "The Impact of the Pull of the Recent on the History of Marine Diversity." *Science* 300 no. 5622 (May 16, 2003). https://doi.org/10.1126/science.1083246.

Jaillarda, Etienne, Peter Bengtsonb, and Annie V. Dhondtc. "Late Cretaceous zmarine Transgressions in Ecuador and Northern Peru: A Refined Stratigraphic Framework." *Journal of South American Earth Sciences* 19 (2005): 307–323. Retrieved January 1, 2003. https://www.researchgate.net/publication/222578015_Late_Cretaceous_marine_transgressions_in_Ecuador_and_northern_Peru_A_refined_stratigraphic_framework.

Jennings, Justin, Félix Palacios Ríos, Nicholas Tripcevich, et al. "The Huarhua Rock Salt Mine: Archaeological Implications of Modern Extraction Practices." UC Berkeley Previously Published Works, April 1, 2013. https://escholarship.org/uc/item/2mt8p588.

King, Hobart M. "Coquina." Geology.com, 2005–2022. https://geology.com/rocks/coquina.shtml.

L'Esperance, Daniel L. "Holzmaden's Jurassic Park: DODDS Students Dig Deep for Fossils." Stars and Stripes, July 17, 2012. https://www.stripes.com/travel/holzmaden-s-jurassic-park-dodds-students-dig-deep-for-fossils-1.183191.

Luzuriaga-Aveiga, Vanessa E., and Diego F. Cisneros-Heredia. "Seasonal Turnover of Avian Community Assembly in a Highly Fragmented Tumbesian Dry Forest of Southwestern Ecuador." *Neotropical Biodiversity* 8, no. 1 (December 31, 2022). https://doi.org/10.1080/23766808.2022.2076784.

Mello, Myrna, Eduardo Koutsoukos, and Erazo Wilmer. "The Napo Formation, Oriente Basin, Ecuador: Hydrocarbon Source Potential and Paleoenvironmental Assessment." Petroleum Source Rocks, January 1995, 167–181. https://www.researchgate.net/publication/295264187_The_Napo_Formation_Oriente_Basin_Ecuador_hydrocarbon_source_potential_and_paleoenvironmental_assessment.

Ministerio de Energia y Minas—Ecuador, eds. "Plan Nacional de Desarrollo del Sector Minero 2020–2030." Ministerio de Energia y Minas—Ecuador, October 2020. https://www.recursosyenergia.gob.ec.

Mönnig, Eckhard, Franz Matthias, and Schweigert Guenter. "The Stratigraphic Chart of Germany (STD 2016): Jurassic." *Zeitschrift der Deutschen Gesellschaft für Geowissenschaften* 169, no. 2 (June 2018). https://doi.org/10.1127/zdgg/2018/0148.

Morante-Carballo, Fernando, Geanella Herrera-Narváez. Nelson Jiménez-Orellana, and Paúl Carrión-Mero. "Puyango, Ecuador Petrified Forest, a Geological Heritage of the Cretaceous Albian-Middle, and Its Relevance for the Sustainable Development of Geotourism." Researchgate.net, August 14, 2020. https://www.researchgate.net/publication/343669250_Puyango_Ecuador_Petrified_Forest_a_Geological_Heritage_of_the_Cretaceous_Albian-Middle_and_Its_Relevance_for_the_Sustainable_Development_of_Geotourism.

Nudds, John, and Paul Selden. "Fossil–Lagerstätten." *Geology Today*, July 22, 2008. https://doi.org/10.1111/j.1365-2451.2008.00679.x.

Prothero, Donald R. *Bringing Fossils to Life: An Introduction to Paleobiology*, 3rd ed. Columbia University Press, 2013.

Ruebsam, Wolfgang, Annette Schmid-Röhl, and Moujahed Al-Husseini. "Astronomical Timescale for the Early Toarcian (Early Jurassic) Posidonia Shale and Global Environmental Changes." *Palaeogeography Palaeoclimatology Palaeoecology* 623, no. 4 (May 2023. https://doi.org/10.1016/j.palaeo.2023.111619.

Sahney, Sarda, Michael J. Benton, and Paul A. Ferry. "Links between Global Taxonomic Diversity,

Ecological Diversity and the Expansion of Vertebrates on land." *Biology Letters* 6, no. 4 (August 23, 2010). https://doi.org/10.1098/rsbl.2009.1024.

Schweitzer, M. H., W. Zheng, T. P. Cleland, et al. "A Role for Iron and Oxygen Chemistry in Preserving Soft Tissues, Cells and Molecules from Deep Time." Proceedings of the Royal Society, November 2013. https://doi.org/10.1098/rspb.2013.2741.

UNESCO.org, eds. "Bosque petrificado de Puyango." *UNESCO.org*, n.d. https://whc.UNESCO.org/en/tentativelists/1081/.

United States Geological Survey, eds. "Examples of Index Fossils." United States Geological Survey, July 28, 1997. http://pubs.usgs.gov/gip/geotime/fossils.html.

Urweltmuseum Hauff, eds. "Collecting Fossils." Urweltmuseum Hauff, N.d. https://www.urwelt-museum.de/en/urweltmuseum/collecting-fossils/.

Wade, Nicholas. "Geological Team Lays Claim to Oldest Known Fossils." *New York Times*, August 21,2011. https://www.nytimes.com/2011/08/22/science/earth/22fossil.html.

## Chapter 6: Calcite and Aragonite

Abbott, A. "Fossil Dealer Charged over Russian Cache." *Nature* 397 (1999). https://doi.org/10.1038/16528.

Alden, Andrew. "Calcite vs Aragonite." ThoughtCo., August 13, 2018. https://www.thoughtco.com/calcite-vs-aragonite-1440962.

Courville, Philippe, and Catherine Crônier. *Diversity or Disparity in the Jurassic (Upper Callovian) Genus Kosmoceras (Ammonitina): Morphometric Approach.* Cambridge University Press, 2016. https://www.cambridge.org/core/journals/journal-of-paleontology/article/abs/diversity-or-disparity-in-the-jurassic-upper-callovian-genus-kosmoceras-ammonitina-a-morphometric-approach/5DBBBEE0469354414230E1A837673B79.

FossilMuseum.net, eds. "Fossils of the Fox Hills Formation." FossilMuseum.net, n.d. http://www.fossilmuseum.net/Fossil_Sites/foxhillsformation.htm.

FossilMuseum.net. "Fossils of the Fox Hills Formation." FossilMuseum.net, n.d. http://www.fossilmuseum.net/Fossil_Sites/foxhillsformation.htm.

Glass, D. J., ed. "Lexicon of Canadian Stratigraphy Volume 4 Western Canada." Canadian Society of Petroleum Geologists, n.d. https://www.scribd.com/document/42826092/Lexicon-4.

"Jurassic Period." World of Earth Science. Encyclopedia.com, August 25, 2022. https://www.encyclopedia.com/science/encyclopedias-almanacs-transcripts-and-maps/jurassic-period.

Li Kui Ming, Edward. "The Most Influential Stone of the Millenium." Korite, December 23, 2013. https://jewelryconnoisseur.net/ammolite/.

Marcinowski, Ryszard, and Jost Wiedmann. "The Albian Ammonites of Poland." *Palaeontologia Polonica*, 35 (1985). https://geojournals.pgi.gov.pl/agp/article/view/9406.

Melroy, Jennifer. "Everything You Need to Know about Fossil Hunting in Badlands National Park, South Dakota." NationalParkObsessed.com, November 30,2020. https://nationalparkobsessed.com/fossil-hunting-in-badlands/.

Mike. "Fox Hills Formation." CSMS Geology Post, May 22, 2012. https://csmsgeologypost.blogspot.com/2012/05/fox-hills-formation.html.

Minnbuckeye. "Fox Hills Ammonites!!" TheFossilForum.com, March 3, 2017. http://www.thefossilforum.com/index.php?/topic/72386-fox-hills-ammonites/.

Paleontica All Things Paleo. "SD Fox Hills Formation." Paleontica All Things Paleo, n.d. https://www.paleontica.org/sites/fossil_site.php?plaats=697&language=en.

Pedersen, Maggie Campbell. *Gem and Ornamental Materials of Organic Origin.* Elsevier, 2003.

Q-files.com, eds. "Jurassic Europe." Q-files.com, 2022. https://www.q-files.com/prehistoric/age-of-dinosaurs/jurassic-europe/print.

Roehler, Henry W. "Stratigraphy of the Upper Cretaceous Fox Hills Sandstone and Adjacent Parts of the Lewis Shale and Lance Formation, East Flank of the Rock Springs Uplift, Southwest Wyoming." U.S. Geological Survey Professional Paper no. 0000. U.S. Government Printing Office (1532), 1993.

Rogov, Mikhail, Baraboshkin, Evgeny Guzhikov, et al. "The Jurassic-Cretaceous Boundary in the Middle Volga Region: Field Guide to the International Meeting on the Jurassic/Cretaceous Boundary." Samara State Technical University, September 7–13, 2015. https://www.researchgate.net/publication/281823637_The_Jurassic-Cretaceous_boundary_in_the_Middle_Volga_region_Field_guide_to_the_International_meeting_on_the_JurassicCretaceous_boundary_September_7-13_2015_Samara_Russia_130_pp.

Tang, C. Marie. "Jurassic Period." *Encyclopedia Britannica*, n.d. https://www.britannica.com/science/Jurassic-Period.

Tellus Storyhive. "Ammolite: Gem of the West." *YouTube.com*, July 15, 2019. https://www.youtube.com/watch?v=jr9kV9OoS_w&t=14s.

## Chapter 7: The Animal in the Ammonite Shell

20minutos.es, eds. "La Unesco declara como reserva de la biosfera al bosque seco del sur de Ecuador." 20 Minutos, January 22, 2015. https://www.20minutos.es/noticia/2354160/0/reserva-biosfera/bosque-seco-ecuador/conservacion-ambiental/.

AEDES, eds. "AEDES—Asociacion Especializada para el Desarrollo Sostenible." La Asociación Especializada para el Desarrollo Sostenible, n.d. http://www.aedes.com.pe/english/index.shtml.

Aquarium of the Pacific, eds. "Ammonite, Subclass Ammonoidea." Long Beach: Aquarium of the Pacific, 2022. https://www.aquariumofpacific.org/onlinelearningcenter/species/ammonite.

Aracari, eds. "The Sillar Route: Why Arequipa Is Called The White City." Aracari, November 8, 2017. https://www.aracari.com/blog/luxury-travel-peru/the-sillar-route-arequipa/.

Arkell, W. J., R. C. Moore, A. K. Miller, W. M. Furnish, and O. H. Schindewolf. *Mesozoic Ammonoidea: Treatise on Invertebrate Paleontology, Part L, Mollusca 4.* Geological Society of American and University of Kansas Press, 1957.

Associated Press, eds. "Peru Concerned about Fossil Trafficking." NBC Universal, February 11, 2007. https://www.nbcnews.com/id/wbna17076190.

Belemniten. "My Little Trip to Solnhofen." *Fossil: Forum*, July 5, 2018. http://www.thefossilforum.com/index.php?/topic/86063-my-little-trip-to-solnhofen/.

BioMed Central Ltd, eds. "Morphological Evolution of Externally Shelled Cephalopods during the Most Intense Phase of the Devonian Nekton Revolution in the Early and Early Middle Devonian." *Springer Nature: BMC Ecology and Evolution* 00, no. 00 (2022): 000–000. https://bmcevolbiol.biomedcentral.com/articles/10.1186/1471-2148-11-115/figures/1.

BioMed Central Ltd, eds. "Parallel Evolution Controlled by Adaptation and Covariation in Ammonoid Cephalopods." *Springer Nature: BMC Ecology and Evolution* 00, no. 00 (2022): 000–000. https://www.bgs.ac.uk/discoveringGeology/time/fossilfocus/images/ammonites/AmmoniteCrossSection.jpg.

Boletzky, S. V. "Cephalopod Development and Evolution." Springer, 1999. https://doi.org/10.1007/978-1-4615-4837-9_23.

Klug, Christian. Paleobiology department member, University of Zürich, Switzerland, Palaeontological Institute and Museum. Email exchange with author, August 10, 2022.

cotahuasiarequipa.com, eds. "Reserva Paisajística de la Subcuenca del Cotahuasi." Portal La Unión, n.d. http://www.cotahuasiarequipa.com/.

Doguzhaeva, Larisa A., Royal H. Mapes, Herbert Summesberger, and Harry Mutvei. *The Preservation of Body Tissues, Shell, and Mandibles in the Ceratitid Ammonoid Austrotrachyceras (Late Triassic), Austria*. Springer, 2007. https://doi.org/10.1007/978-1-4020-6806-5_11.

Emling, Shelley. *The Fossil Hunter: Dinosaurs, Evolution, and the Woman Whose Discoveries Changed the World*. Palgrave Macmillan, 2009.

Eyden, Phil. "Ammonites: A General Overview." Tonmo.com, March 9, 2019. https://www.tonmo.com/pages/ammonites/.

Flower, Rousseau H. "Nautiloid Shell Morphology." New Mexico Bureau of Mines and Mineral Resources, Memoir. 13, 1964.

Instituto de Investigación Geológico y Energético Gobierno del la República del Ecuador. "Bosque Petrificado de Puyango camino a ser el segundo Geoparque Mundial del Ecuador." Instituto de Investigación Geológico y Energético, N.d. https://www.geoenergia.gob.ec/bosque-petrificado-de-puyango-camino-a-ser-el-segundo-geoparque-mundial-del-ecuador/.

Jaillarda, Etienne, Peter Bengtsonb, and Annie V. Dhondtc. "Late Cretaceous Marine Transgressions in Ecuador and Northern Peru: A Refined Stratigraphic Framework." *Journal of South American Earth Sciences* 19, no. 3 (August 2005): 307–323. https://doi.org/10.1016/j.jsames.2005.01.006.

Jennings, Justin, Félix Palacios Ríos, Nicholas Tripcevich, et al. *The Huarhua Rock Salt Mine: Archaeological Implications of Modern Extraction Practices*. University of California, Berkeley, 2001. https://escholarship.org/uc/item/2mt8p588.

Jurassic Coast World Heritage Site Partnership, eds. "Ammonites: The Ultimate Jet-Propelled shells!" Jurassic Coast World Heritage Site Partnership, 2022. Sttps://jurassiccoast.org/fossil-favourites/ammonite/.

Keupp, H., R. Hoffmann, K. Stevens, and R. Albersdörfer. "Key Innovations in Mesozoic Ammonoids: The Multicuspidate Radula and the Calcified Aptychus." *Palaeontology* 59, no. 6 (2016): 775–791.

King, Andy H., and David H. Evans. "High-Level Classification of the Nautiloid Cephalopods: A Proposal for the Revision of the Treatise Part K." *Swiss Journal of Palaeontology* 138 no. 1 (2019). https://doi.org/10.1007/s13358-019-00186-4.

King, Hobart M. "Coquina: A Porous Limestone Composed Almost Entirely of Fossil Debris." Geology.com, 2022. https://geology.com/rocks/coquina.shtml.

Klug, Christian, et al., eds. *Ammonoid Paleobiology: From Anatomy to Ecology, Topics in Geobiology 43*. Springer Science + Business Media, 2015.

Klug, Christian, Kenneth De Baets, Björn Kröger, Mark A. Bell, Dieter Korn, and Jonathan L. Payne. "Normal Giants? Temporal and Latitudinal Shifts of Palaeozoic Marine Invertebrate Gigantism and Global Change." *Lethaia* 48, no. 2 (2015). https://doi.org/10.1111/let.12104.

Klug, Christian, and Jens Lehmann. "Soft Part Anatomy of Ammonoids: Reconstructing the Animal Based on Exceptionally Preserved Specimens and Actualistic Comparisons." Ammonoid Paleobiology: From Anatomy to Ecology. Springer Science + Business Media, 2015, 507–529. http://coo.fieldofscience.com/2017/02/the-arms-of-ammonite.html.

Klug, Christian, Wolfgang Riegraf, and Jens Lehmann. "Soft–Part Preservation in Heteromorph Ammonites from the Cenomanian–Turonian Boundary Event (OAE 2) in North–West Germany." *Palaeontology* 55, no. 6 (December 2012): 1307–1331. https://doi.org/https://doi.org/10.1111/j.1475-4983.2012.01196.x.

Klug, Christian, Guenter Schweigert, and Helmut Tischlinger. "Failed Prey or Peculiar Necrolysis? Isolated Ammonite Soft Body from the Late Jurassic of Eichstätt (Germany) with Complete Digestive Tract and Male Reproductive Organs."

*Swiss Journal of Palaeontology* 00, no. 00 (January 2021): 000–000. https://doi.org/10.1186/s13358-020-00215-7.

Kröger, Björn. *The Size of the Siphuncle in Cephalopod Evolution*. Senckenbergiana Lethaea, 2003. https://doi.org/10.1007/BF03043304.

Kröger, Björn, and Ed Landing. *Onset of the Ordovician Cephalopod Radiation: Evidence from the Rochdale Formation (Middle Early Ordovician, Stairsian) in Eastern New York*. Cambridge University Press, 2008. https://doi.org/10.1017/S0016756808004585.

Kröger, Björn, and Yun-Bai, Zhang. "Pulsed Cephalopod Diversification during the Ordovician." *Palaeogeography, Palaeoclimatology, Palaeoecology* 273 (2008). https://doi.org/10.1016/j.palaeo.2008.12.015.

Lindgren, A. R., G. Giribet, and M. K. Nishiguchi. "A Combined Approach to the Phylogeny of Cephalopoda (Mollusca)." *Cladistics* 20, no. 5 (October 2004). https://doi.org/10.1111/j.1096-0031.2004.00032.x.

Marcinowski, Ryszard, and Jost Wiedmann. "The Albian Ammonites of Poland." *Palaeontologia Polonica* 35 (1985). https://geojournals.pgi.gov.pl/agp/article/view/9406.

McAfee, Kathleen. "Cotahuasi Canyon: Explore the World's 3rd Deepest Canyon." Peru for Less LLC, June 2, 2014. https://www.peruforless.com/blog/cotahuasi-canyon/.

McGowan, Christopher. *The Dragon Seekers*. Perseus, 2001.

McKeever, Amy. "What Are Ammonites, and How Did They Come to Rule the Prehistoric Seas?" NationalGeographic.com, Month 00, 000. https://www.nationalgeographic.com/animals/facts/ammonites.

Mello, M. R., E. A. M. Koutsoukos, and W. Z. Erazo. *The Napo Formation, Oriente Basin, Ecuador: Hydrocarbon Source Potential and Paleoenvironmental Assessment*. Petroleum Source Rocks,

Casebooks in Earth Sciences. Springer, 1995. https://doi.org/10.1007/978-3-642-78911-3_10.

Monks, Neale. "Ammonite Wars." Deposits Mag, February 25, 2016. https://depositsmag.com/2016/02/25/ammonite-wars/.

Monks, Neale, and Philip Palmer. *Ammonites*. Natural History Museum, 2002.

Monks, Neale and Jeremy R. Young. "Body Position and the Functional Morphology of Cretaceous Heteromorph Ammonites," *Palaeontologia Electronica* 1, no. 1; (1998). https://doi.org/10.26879/98001.

Monnet, C., K. De Baets, and C. Klug. "Parallel Evolution Controlled by Adaptation and Covariation in Ammonoid Cephalopods." *BMC Evolutionary Biology* 11, no. 115 (2011). https://doi.org/10.1186/1471-2148-11-115.

Monterey Bay Aquarium, eds. "Animals A to Z: Cephalopods." Monterey Bay Aquarium, n.d. https://www.montereybayaquarium.org/animals/animals-a-to-z/cephalopods.

Morante-Carballo, Fernando. Geanella Herrera-Narváez, Nelson Jiménez-Orellana, and Paúl Carrión-Mero. *Puyango, Ecuador Petrified Forest, a Geological Heritage of the Cretaceous Albian-Middle, and Its Relevance for the Sustainable Development of Geotourism*. Centro de Investigación y Proyectos Aplicados a las Ciencias de la Tierra (CIPAT), ESPOL Polytechnic, 2020.

Mutvei, Harry; Yun-bai Zhang, and Elena Dunca. "Late Cambrian Plectronocerid Nautiloids and Their Role in Cephalopod Evolution." *Palaeontology* 50, no. 6 (2007). https://doi.org/10.1111/j.1475-4983.2007.00708.x.

Nautiloid Network, The, eds. "Solnhofen, Germany." Nautiloid Network, n.d. http://nautiloid.net/fossils/sites/solnhofen/solnhofen.html.

Norman, Mark. *Cephalopods: A World Guide*. ConchBooks, 2000.

Olóriz Sâez, Federico, and Francisco J. Rodriguez-Tovar III. "Advancing Research on Living and

Fossil Cephalopods: Development and Evolution: Form, Construction, and Function: Taphonomy, Palaeoecology, Palaeobiogeography, Biostratigraphy, and Basin Analysis." Fourth International Symposium on Cephalopods: Present and Past, Granada, Spain, 1996. Springer Science+Business Media. https://doi.org/10.1007/978-1-4615-4837-9.

Olóriz, Federico, and Francisco J. Rodríguez-Tovar, eds. *Advancing Research on Living and Fossil Cephalopods*. Springer, 1999. https://doi.org/10.1007/978-1-4615-4837-9_1.

Oz Fossils, eds. "The Age of Reptiles." Oz Fossils, ABC Online, 2020. https://www.abc.net.au/science/ozfossil/ageofreptiles/eromanga/ammonites.htm.

painshill. "Did Ammonites Have Ink Sacks?" *Fossil Forum*, July 26, 2014. http://www.thefossilforum.com/index.php?/topic/48333-did-ammonites-have-ink-sacks/.

Pohle, Alexander, Björn Kröger, Rachel C. M. Warnock, et al. "Early Cephalopod Evolution Clarified through Bayesian Phylogenetic Inference." *BMC Biology* 20, no. 88 (2022). https://doi.org/10.1186/s12915-022-01284-5.

Quora.com, eds. "How Many Tentacles Does a Nautilus Have?" Quora.com, N.d. https://www.quora.com/How-many-tentacles-does-a-nautilus-have.

Robert, Emmanuel, and Luc Georges Bulot. "The Albian Ammonites Successions of the Subandean Zone of Ecuador." European Meeting on the Palaeontology and Stratigraphy of Latin America. Instituto Geológico y Minero de España, 2007.

Romero, Simon, and Andrea Zarate. "Peru's Ocucaje Desert Attracts Fossil Hunters and Smugglers." *New York Times*, December 11, 2010. https://www.nytimes.com/2010/12/12/world/americas/12peru.html.

Rowe, Alison J., Neil H. Landman, J. Kirk Cochran, James D. Witts, and Matthew P. Garb. "Late

Cretaceous Methane Seeps as Habitats for Newly Hatched Ammonites." *Palaios* 35, no. 3 (March 26, 2020). https://doi.org/10.2110/palo.2019.105.

Sandoval, Carla. "La Unesco declara como reserva de la biosfera al bosque seco del sur de Ecuador." Grupo El Comercio, January 21, 2015. https://www.ambiente.gob.ec/ecuador-tiene-una-nueva-reserva-de-biosfera-reconocida-por-unesco/.

Sarti, Carlo. "Whorl Width in the Body Chamber of Ammonites as a Sign of Dimorphism." Advancing Research on Living and Fossil Cephalopods. Springer, 1999. https://doi.org/10.1007/978-1-4615-4837-9_23.

Smith, C. P. A., N. H. Landman, J. Bardin, and I. Kruta. "New Evidence from Exceptionally 'Well-Preserved' Specimens Sheds Light on the Structure of the Ammonite Brachial Crown." *Scientific Reports* 11, no. 11862 (June 4, 2021). https://doi.org/10.1038/s41598-021-89998-4.

Stehli, F. G. "Shell Mineralogy in Paleozoic Invertebrates." *Science* 123, no. 3206 (June 1956). https://doi.org/10.1126/science.123.3206.1031.

Stridsberg, Sven. "Aptychopsid Plates: Jaw Elements or Protective Operculum." Lethaia 17, no. 1 (January 1984). https://doi.org/10.1111/j.1502-3931.1984.tb00670.x.

Tanabe, K., I. Kruta, and N. H. Landman. 2015. "Ammonoid Buccal Mass and Jaw Apparatus." Ammonoid Paleobiology: From Anatomy to Ecology. Springer Science + Business Media, 2015, 429–484. https://doi.org/10.1007/978-94-017-9630-9_10.

Tripadvisor, eds. "Canon de Cotahuasi." Tripadvisor, n.d. https://www.tripadvisor.in/ShowUserReviews-g294313-d7161690-r632925047-Canon_de_Cotahuasi-Arequipa_Arequipa_Region.html.

Tripadvisor, Germany Travel Forum Editors. "Fossil Hunting in Bavaria." Tripadvisor, 2013. https://www.tripadvisor.com/ShowTopic-g187275-i116-k6446560-Fossil_Hunting_in_Bavaria-Germany.html.

Tripadvisor, Peru Travel Forum Editors. "Cotahuasi Is One of the Delights of the World." Tripadvisor, July 2016. https://www.tripadvisor.com/Location-PhotoDirectLink-g5616269-d9998196-i248587732-Hotel_Vallehermoso-Cotahuasi_Arequipa_Region.html.

Tripadvisor, Peru Travel Forum Editors. "Warning about Fossils." Tripadvisor, 2010. https://www.tripadvisor.com/ShowTopic-g294311-i818-k3232984-Warning_about_fossils-Peru.html.

UK Research and Innovation Editors. "Neutrons Reveal Ammonites Jetted around Prehistoric Oceans." UK Research and Innovation, December 17, 2021. https://www.ukri.org/news/neutrons-reveal-ammonites-jetted-around-prehistoric-oceans/.

UNEP-WCMC. "Protected Area Profile for Subcuenca del Cotahuasi." World Database on Protected Areas, October 2023. www.protectedplanet.net.

UNESCO.org, eds. "Bosque petrificado de Puyango." UNESCO.org, n.d. https://whc.UNESCO.org/en/tentativelists/1081/.

UNESCO World Heritage Convention Editors. "Landscape Reserve Sub Cuenca del Cotahuasi." UNESCO.org, n.d. https://whc.unesco.org/en/tentativelists/6423/.

Von Byern, Janek, and Waltraud Klepal. "Adhesive Mechanisms in Cephalopods: A Review." *Biofouling* (February 2006). https://doi.org/10.1080/08927010600967840.

Walla, G. "A Study of the Comparative Morphology of Cephalopod Armature." tonmo.com, 2007.

Zych, Ariel. "Jet-Setting Cephalopods." Sciencefriday.com, June 19, 2015. https://www.sciencefriday.com/educational-resources/jet-setting-cephalopods/.

## Chapter 8: Size Matters

Bert, Didier, Jean-Simon Pagès. "Overview of the Les Isnards Ammonite Slab Geoheritage Site (Digne-les-Bains, Southeastern France)." Islamic Azad University-Isfahan Branch. Geoconservation Research, September 1, 2021. https://gcr.isfahan.iau.ir/article_686264.html.

Communauté de Communes du Val d'Argent, eds. "Partez à la découverte des mines d'argent." Communauté de Communes du Val d'Argent, 2014. http://capitale-mineralogie.fr/visite_mines/.

Communauté de Communes du Val d'Argent, eds. "Tourisme en Val d'Argent." Communauté de Communes du Val d'Argent, 2014. http://capitale-mineralogie.fr/.

De Baets, Kenneth, Christian Klug, Dieter Korn, and Neil H. Landman. "Early Evolutionary Trends in Ammonoid Embryonic Development." *International Journal of Organic Evolution* (February 14, 2012). https://doi.org/10.1111/j.1558-5646.2011.01567.x.

EUROMIN Project, The, eds. "The Mines of Sainte-Marie-aux-Mines." EUROMIN Project, n.d. http://euromin.w3sites.net/Nouveau_site/gisements/StMarie/GISMARe.htm#1.

Everything Dinosaur, eds. "Ammonites Separating the Boys from the Girls." Everything Dinosaur, 2022. https://blog.everythingdinosaur.com/blog/_archives/2019/06/09/ammonites-separating-the-boys-from-the-girls.html.

Eyden, Phil. "Ammonites: A General Overview." TONMO, March 9, 2019. https://www.tonmo.com/pages/ammonites/.

Ganguly, T., and S. Bardhan. "Dimorphism in Placenticeras mintoi from the Upper Cretaceous Bagh Beds, Central India." *Cretaceous Research* 14, no. 6 (1993): 747–756. https://doi.org/10.1006/cres.1993.1050.

Hesselbo, S. P., J. G. Ogg, M. Ruhl, L. A. Hinnov, and C. J. Huang. "Chapter 26—The Jurassic Period, Geologic Time Scale, 2020." Elsevier, 2020, 955–1021, https://doi.org/10.1016/B978-0-12-824360-2.00026-7.

Lithologia.com, eds. "Fossils for Sale We Are Digging: French Triassic Upper Muschelkalk." Lithologia.

com, n.d. https://www.lithologia.com/Fossils-for
-sale-we-are-digging-french-Triassic-upper
-Muschelkalk-bchaaaaaa.asp.

Mineral & Gem, eds. "Rendez-vous en 2023! 58ème
Mineral & Gem." Mineral & Gem | Propulsé
par APPOE, N.d. www.sainte-marie-mineral
.com.

Paigés, J.-S. "The GeoPark of Haute-Provence,
France—Geology and Palaeontology Protected
for Sustainable Development, PaleoParks—The
Protection and Conservation of Fossil Sites
Worldwide." Carnets de Géologie/Notebooks on
Geology, 2009.

Parent, H., and Zatoń, M. "Sexual Dimorphism in
the Bathonian Morphoceratid Ammonite Poly-
sphinctites tenuiplicatus." Acta Palaeontologica
Polonica 61, no. 4 (2016): 875–884.

Plotnick, Roy E. "Remarks." Paleontologica Electronica 12, no. 1 (2009). Web version; 1532–3056.

Shepherd, Roy. "What Is an Ammonite?" Discovering Fossils, 2022. http://www.discoveringfossils
.co.uk/ammonites.htm.

UNESCO Géoparc de Haute-Provence. "Welcome
in the Haute Provence Geopark." UNESCO
Géoparc de Haute-Provence, n.d. https://www
.geoparchauteprovence.com/uk/s564461993
.siteweb-initial.fr/index.html.

University of Zürich. "First Fertile, Then Futile:
Ammonites Change in Reproductive Strategy
Helped Them Survive Three Mass Extinctions."
ScienceDaily, n.d. www.sciencedaily.com/releases
/2012/04/120423104522.htm.

Val d'Argent, Tourism Office, eds. "Val d'Argent—
Tourist Information." Communication Agency
Alsace Greencub, n.d. https://www.valdargent
-tourisme.fr/en.

Visit Alsace, eds. "Alsace International Mineral &
Gem Dhow in Sainte-Marie-aux-Mines." Thu-
ria, n.d. https://www.visit.alsace/en/228001858
-international-mineral-gem-show-in-sainte-marie
-aux-mines/.

Wikipedia, eds. "Ammonoidea." Wikipedia.com,
September 2004. https://en.wikipedia.org/wiki
/Ammonoidea.

Wikipedia, eds. "Arietitidae." Wikipedia.com, October
2010. https://en.wikipedia.org/wiki/Arietitidae.

Wikipedia, eds. "Bactritida." Wikipedia.com, February
14, 2006. https://en.wikipedia.org/wiki/Bactritida.

Wikipedia, eds. "Cephalopod egg fossil." Wikipedia.
com, June 2013. https://en.wikipedia.org/wiki
/Cephalopod_egg_fossil.

Wikipedia, eds. "Muschelkalk." Wikipedia.com, October
2005. https://en.wikipedia.org/wiki/Muschelkalk.

Wikipedia, eds. "Sainte-Marie-aux-Mines." Wikipe-
dia.com, November 2005. https://en.wikipedia
.org/wiki/Sainte-Marie-aux-Mines.

Wikipedia, eds. "Schlotheimiidae." Wikipedia.
com, June 2014. https://en.wikipedia.org/wiki
/Schlotheimiidae.

Wikipedia, eds. "UNESCO Global Geoparks." Wiki-
pedia.com, September 2009. https://en.wikipedia
.org/wiki/UNESCO_Global_Geoparks.

YvesProvence. "La Dalle au Ammonites—Digne les
Bains—Paleontology and Fossils on Waymarking.
com." Waymarking.com, July 23, 2017. https://
www.waymarking.com/waymarks/WMW8AA
_La_Dalle_au_Ammonites_Digne_les_Bains.

## Chapter 9: What Ammonites Ate

ABC.net.au, eds. "The Age of Fossils: Ammonites."
ABC, 2020. https://www.abc.net.au/science
/ozfossil/ageofreptiles/eromanga/ammonites.htm.

Andalucia.com, eds. "El Torcal de Antequera." About
Andalucia.com, n.d. https://www.andalucia.com
/antequera/torcal/home.htm.

Aquarium of the Pacific, eds. "Ammonite: Subclass Ammonoidea." Aquarium of the Pacific,
2022. http://www.aquariumofpacific.org/online
learningcenter/species/ammonite.

Arkell, W. J., W. M. Furnish, Bernhard Kummel, et
al. Treatise on Invertebrate Paleontology, Part L,

*Mollusca 4.* Geological Society of America and University of Kansas Press, 1957.

asniper8541. "Fossil Hunting Trips: Lisbon Portugal." *Fossil Forum*, September 1, 2018. http://www.thefossilforum.com/index.php?/topic/88000-lisbon-portugal/.

Bergmann, Lothar. "Arte sureño: El arte rupestre del extremo sur de la Península Ibérica." AGEDPA: Asociación Gaditana para el Estudio y la Defensa del Patrimonio Arqueológico, 2009. http://www.arte-sur.com/.

Billert, Thomas. "Fossils on Majorca: The Collection José Juárez Ruiz and an Excursion to Majorcan Lower Cretaceous." Steinkern.de, August 16, 2013. https://www.steinkern.de/english-articles/959-fossils-on-majorca-the-collection jose-jua-rez-ruiz-and-majorcan-lower-cretaceous.html.

Borges, Marisa E. N., James B. Riding, Paulo Fernandes, and Zélia Pereira. "Callovian (Middle Jurassic) Dinoflagellate Cysts from the Algarve Basin, Southern Portugal." *Review of Palaeobotany and Palynology* 170 (2012): 40–56. https://doi.org/10.1016/j.revpalbo.2011.10.008.

Catalán, Manuel Pellicer. "Las culturas del neolítico-calcolítico en Andalucía Oriental." Espacio Tiempo y Forma Serie I Prehistoria y Arqueología, January 1995. https://doi.org/10.5944/etfi.8.1995.4620.

Chaplow, Chris. "The Antequera Dolmen Site—Dólmenes de Antequera—Conjunto Arqueológico Menga, Viera and El Romeral Dolmens." About Andalucia.com, N.d. https://www.andalucia.com/antequera/dolmens-de-menga.htm.

Costatropical.net, eds. "El Torcal de Antequera Top 2022 Guide!" Costatropical.net, 2022. https://www.costatropical.net/el-torcal-de-antequera/.

Davis, Danny Lee. "Commercial Navigation in the Greek and Roman World." University of Texas at Austin, May 2009. https://www.academia.edu/2955208/Dissertation_Univ._of_Texas_at_Austin_Commercial_Navigation_in_the_Greek_and_Roman_World.

Dracontes. "Finding Fossils in Portugal." *Reddit.com*, November 2020. https://www.reddit.com/r/Paleontology/comments/np6ez7/finding_fossils_in_portugal/?rdt=39508.

Dommergues, Jean-Louis, Christian Meister, and Rogerio B. Rocha. "The Pliensbachian Ammonites of the Algarve Basin (Portugal) and Their Palaeobiogeographical Significance for the 'Iberia-Newfoundland' Conjugate Margins." Swiss Geological Society 104 (March 18, 2011): 81–96. https://doi.org/10.1007/s00015-011-0056-2.

Educared.net, eds. "ENTORNO:" *Educared.net*, October 29,2008. http://www.educared.net/concurso2001/663/entorno.htm.

Eyden, Phil. "Ammonites: A General Overview." Deep Intuition LLC, March 9, 2019. https://www.tonmo.com/pages/ammonites/.

Fernández-López, Sixto R., Maria Helena Henriques, and Charles Mangold. "Ammonite Succession at the Bajocian/Bathonian Boundary in the Cabo Mondego Region (Portugal)." *Lethaia* 39, no. 3 (March 12, 2007). https://doi.org/10.1080/00241160600847405.

Fossil Huntress. Facebook Post. *Facebook*, February 18, 2021. https://www.facebook.com/permalink.php?id=128899300401&story_fbid=10164742550150402.

Fossil Huntress. "Pliensbachian: Apoderoceras." Fossil Huntress, January 11, 2020. https://fossilhuntress.blogspot.com/2020/01/sexual-dimorphism-apoderoceras.html.

Fuerte Hoteles, eds. "We Give You 11 Reasons Why You Should Not Miss El Torcal de Antequera. The Most Original Natural Place of Malaga." Fuerte Group Hotels, October 13, 2020. https://blog.fuertehoteles.com/en/activities/visit-the-torcal-de-antequera/.

Gesch, Lori. "Whew . . . It's Finally Out!" Gesch What, 2022. http://www.areallycrappystory.com/.

Giornale di scienza, Galileo. "Ecco i denti delle ammo-niti.mpg." *YouTube.com*, January 7, 2011. https://www.youtube.com/watch?v=as4yLMj_pgo.

Gómez, Caro, and José Antonio. "Yacimientos e industrias achelenses en las terrazas fluviales de la depresión del bajo Guadalquivir (Andalucía, España). Secuencia estratigráfica, caracterización tecnocultural y cronología." *Revista de estudios locales* 4, no. 00 (2006) 000–000.

Hunter0811. "Exporting Fossils from Portugal." *Fossil Forum*, November 20, 2019. http://www.thefossilforum.com/topic/100133-exporting-fossils-from-portugal/.

IFAR, eds. "Country Summary for SPAIN." International Foundation for Art Research (IFAR), n.d. https://www.ifar.org/country_title.php?docid=1217632312.

Johannes. "Is Fossil Collecting Legal in Spain?" *Fossil Forum*, June 16, 2016. http://www.thefossilforum.com/topic/66034-is-fossil-collecting-legal-in-spain/.

Kruta, Isabelle, Neil Landman, Isabelle Rouget, Fabrizio Cecca, and Paul Tafforeau. "The Role of Ammonites in the Mesozoic Marine Food Web Revealed by Jaw Preservation." *Science* 331, no. 6013 (January 2011). https://doi.org/10.1126/science.1198793.

Málaga Adventures, eds. "A Guide to El Torcal de Antequera." Málaga Adventures, 2019. https://malagaadventures.com/el-torcal-de-antequera/.

Marshall, Michael. "Ammonites' Strict Diet Doomed Them to Extinction." New Scientist Ltd., January 6, 2011. https://www.newscientist.com/article/dn19924-ammonites-strict-diet-doomed-them-to-extinction/#ixzz6SVFETr00.

Marzo, Mariano, Violeta Borruel-Abadía, José López-Gómez, et al. "Climate Changes during the Early–Middle Triassic Transition in the E. Iberian Plate and their Palaeogeographic Significance in the Western Tethys Continental Domain." *Palaeogeography, Palaeoclimatology, Palaeoecology* 00, no. 00 (December 15, 2015): 000–000. https://doi.org/10.1016/j.palaeo.2015.09.043.

Marzola, Marco, Octávio Mateus, and Miguel Moreno-Azanza. "Field Trip: The Paleontology of Algarve 2 Days (June 30–July 01)." Conference: XVI Annual Meeting of the European Association of Vertebrate Palaeontology, Caparica, Portugal. Universidade Nova de Lisboa, 2022. https://eventos.fct.unl.pt/eavp2018/pages/fieldtrips.

Matamales-Andreu, Rafel, Enrique Peñalver, Eudald Mujal, et al. "Early–Middle Triassic Fluvial Ecosystems of Mallorca (Balearic Islands): Biotic Communities and Environmental Evolution in the Equatorial Western Peri-Tethys." *Earth-Science Reviews* 222 (2021). https://doi.org/10.1016/j.earscirev.2021.103783.

Oliveira, Jose Tomas, Paulo Fernandes, Zelia Pereira, and Marisa E. N. Borges. "The Telheiro Beach Variscan Unconformity and Mesozoic Algarve Basin Sections." LNEG, September 2009. https://www.researchgate.net/publication/289130086.

Olóriz, F., and F. J. Rodríguez-Tovar. "The Ammonite Sutneria from the Upper Jurassic of Southern Spain." *Palaeontology* 39, no. 4 (1996): 851–867. https://www.palass.org/publications/palaeontologyjournal/archive/39/4/article_pp851-867.

Paleontica.org, eds. "Algarve." Stichting Paleontica, n.d. https://www.paleontica.org/sites/fossil_site.php?plaats=129.

Paleontica.org, eds. "Mallorca." Stichting Paleontica, n.d. https://www.paleontica.org/locations/fossil/210.

Paleozoo.com.au, eds. "Ammonites." Paleozoo.com.au, n.d. http://www.paleozoo.com.au/Ammonite.php.

Parque Natural Torcal de Antiquera, eds. "Route of the Ammonites." Parque Natural Torcal de Antiquera, n.d. http://www.torcaldeantequera.com/en/guided-tours/guided-visits/ammonites-path/.

Pohle, Alexander, Dirk Fuchs, Dieter Korn, and Christian Klug. "Spatial Distribution of Oncocerids Cephalopods on a Middle Devonian Bedding Plane Suggests Semelparous Life Cycle." Scientific Reports 10, no. 1 (February 2020): 2847. https://doi.org/10.1038/s41598-020-59507-0 www.nature.com/scientificreports.

Scheffel, Richard L., and Susan J. Wernet. "Natural Wonders of the World. United States of America." Reader's Digest Association, 1980.

Segura, M., Fernando Barroso-Barcenilla, Pedro M. Callapez, and Javier Gil. "Depositional Sequences and Cephalopod Assemblages in the Upper Cenomanian–Lower Santonian of the Iberian Peninsula (Spain and Portugal)." Geologica Acta 37, no. 1 (March 2014). https://doi.org/10.1344/105.000002056.

Shareit. "Hands Off Our Valuable Fossils! Portugal's Paleontology 'Up for Grabs.'" Portugal News, February 1, 2018. https://www.portugalresident.com/hands-off our-valuable-fossils-portugals-paleontology-up-for-grabs/.

Shepherd, Roy. "What is an Ammonite?" Discoveringfossils.co.uk, n.d. http://www.discoveringfossils.co.uk/ammonites.htm.

Storetvedt, K. M. "The Tethys Sea and the Alpine-Himalayan Orogenic Belt: Mega-Elements in a New Global Tectonic System." Physics of the Earth and Planetary Interiors 62, no. 1–2 (1990). https://doi.org/10.1016/0031-9201(90)90198-7.

Suess, Eduard, W. J. Sollas, and Hertha Sollas, eds. The Face of the Earth [Das Antlitz der Erde]. Vol. I. Clarendon, 1904.

Switek, Brian. "Ammonite Eats." Wired.com, April 29, 2012. https://www.wired.com/2012/04/ammonite-eats/.

Tavera, J. M., A. Checa, F. Olóriz, and M. Company. "Mediterranean Ammonites and the Jurassic-Cretaceous Boundary in Southern Spain (Subbetic Zone)." University of Granada, January 1986. https://www.researchgate.net/publication/313166851.

Torcal de Antequera, eds. "Guided and Free Tours." Torcal de Antequera, 2022. https://www.torcaldeantequera.com/en/guided-tours/.

Torcal de Antequera, eds. "Wildlife and Vegetation." Torcal de Antequera, 2022. https://www.torcaldeantequera.com/en/el-torcal/wildlife-and-vegetation/.

UNESCO World Heritage Convention, eds. "Antequera Dolmens Site." UNESCO World Heritage Convention, n.d. https://whc.unesco.org/en/list/1501/.

UNESCO World Heritage Convention, eds. "Southwest Coast." UNESCO World Heritage Convention, n.d. https://whc.unesco.org/en/tentativelists/6225/.

USGS, eds. "Where Do Earthquakes Occur?" U.S. Geological Survey (USGS), n.d. https://www.usgs.gov/faqs/where-do-earthquakes-occur.

Vallespí Pérez, Enrique. "Las industrias achelenses de Andalucía: ordenación y comentarios." SPAL: Revista de Prehistoria y Arqueología de la Universidad de Sevilla 1 (1992). https://doi.org/10.12795/spal.1992.i1.03.

Vega, Toscano, Luis Gerardo. "La Cueva de la Carihuela, un enclave básico situado en la localidad de Píñar (Granada)." Waste Magazine, n.d. https://wastemagazine.es/primeuro19.htm.

Wikisource.org, eds. "Page: Quarterly Journal of the Geological Society of London, vol. 27.djvu/227." Wikisource.org, August 8. 2017. https://en.wikisource.org/wiki/Page:Quarterly_Journal_of_the_Geological_Society_of_London,_vol._27.djvu/227.

## Chapter 10: What Ate Ammonites

andrewb1994. "Fossils in Switzerland." Fossil Forum, March 3, 2012. http://www.thefossilforum.com/index.php?/topic/27963-fossils-in-switzerland/.

Balini, Marco, Bogdan Jurkovsek, and Tea Kolar-Jurkovšek. "New Ladinian Ammonoids from MT. Svilaja (External Dinarides, Croatia)." Rivista

*Italiana di Paleontologia e Stratigrafia* (November 2006). https://doi.org/10.13130/2039-4942/6348.

Beltrán-Triviño, Alejandro, Wilfried Winkler, Albrecht von Quadt, and Daniela Gallhofer. "Triassic Magmatism on the Transition from Variscan to Alpine Cycles: Evidence from U–Pb, Hf, and Geochemistry of Detrital Minerals." *Swiss Journal of Geosciences* (2016). https://doi.org/10.1007/s00015-016-0234-3.

Bernasconi, Stefano Michele. "Geochemical and Microbial Controls on Dolomite Formation and Organic Matter Production/Preservation in Anoxic Environments: A Case Study from the Middle Triassic Grenzbitumenzone, Southern Alps (Ticino, Switzerland)." PhD diss., ETH Zürich, 1991. https://doi.org/10.3929/ethz-a-000611458.

Clarkson, E. N. K. *Invertebrate Palaeontology and Evolution*, 4th ed. Wiley-Blackwell, 1998.

Convenzione Italiana dei Comuni, eds. "Welcome to the Monte San Giorgio UNESCO's World Heritage Site." Convenzione Italiana dei Comuni, N.d. http://www.montesangiorgio.org/.

Dietze, Volker, Stefano Cresta, Luca Martire, and Giulio Pavia. "Ammonites, Taphonomical Analysis and Biostratigraphy of the Aalenian (Middle Jurassic) from Capo San Vigilio (Lake Garda, N Italy)." *Schweizerbart Science* 284, no 2 (May 2017): 161–206. https://doi.org/10.1127/njgpa/2017/0657.

*Economist, The*, eds. "Remember the Tooth: A Link Is Made in an Ancient Ecosystem." *The Economist*, April 16, 2011. https://www.economist.com/science-and-technology/2011/04/14/remember-the-tooth.

Edwards, Lin. "Ammonites Were Probably Eaten by Fellow Cephalopods." *Phys.org*, December 3, 2010. https://phys.org/news/2010-12-ammonites-eaten-fellow-cephalopods.html.

Federal Food Safety and Veterinary Office, eds. "Souvenirs." Federal Food Safety and Veterinary Office (FSVO), n.d. https://www.blv.admin.ch/blv/en/home/gebrauchsgegenstaende/reisen-mit-gebrauchsgegenstaenden/souvenirs.html.

Hall, Danielle. "Meals of the Mighty Mosasaur." Smithsonian Institution, November 2018. https://ocean.si.edu/through-time/ancient-seas/meals-mighty-mosasaur.

Italy Heritage.com, eds. "Monte San Giorgio, Province of Varese, Lombardy, Italy." Italy Heritage, 2022. https://www.enchantingitaly.com/landmarks/lombardia/varese/monte-san-giorgio.htm.

Klug, Christian. Department of palaeontology, University of Zürich, Switzerland, Palaeontological Institute and Museum. Email exchange with author, November 30, 2023.

Klug, Christian, Guenter Schweigert, Helmut Tischlinger, and Helmut Pochmann "Failed Prey or Peculiar Necrolysis? Isolated Ammonite Soft Body from the Late Jurassic of Eichstätt (Germany) with Complete Digestive Tract and Male Reproductive Organs." *Swiss Journal of Palaeontology* 140, no. 1 (January 2021). https://doi.org/10.1186/s13358-020-00215-7.

Kowinsky, Jayson. "Mosasaur: The Great Marine Reptile." *Fossilguy.com*, 2022. https://www.fossilguy.com/gallery/vert/reptile/mosasaur/index.htm.

Mietto, Paolo, Stefano Manfrin, Nereo Preto, and Piero Gianolla. "Selected Ammonoid Fauna from Prati di Stuores/Stuores Wiesen and Related Sections across the Ladinian/Carnian Boundary (Southern Alps, Italy)." *Rivista Italiana de Paleontologia e Stratigrafia* (November 2008). https://doi.org/10.13130/2039-4942/5909.

Ministry for Cultural Heritage and Activities, "Legislative Decree no. 42 of 22 January 2004: Code of the Cultural and Landscape Heritage." UNESCO World Heritage Centre, June 2004. https://whc.unesco.org/document/155711.

Monks, Neale, and Phil Palmer. *Ammonites*. Smithsonian Books, 2002.

MySwitzerland.com, eds. "Monte San Giorgio." Switzerland Tourism, 2022. https://www

.myswitzerland.com/en-us/experiences /monte-san-giorgio/.

Ogg, J. G. "Chapter 25: Triassic." In *The Geologic Time*, 681–730. Elsevier B.V., 2012. https://doi .org/10.1016/B978-0-444-59425-9.00025–1.

Rieppel, Olivier. *Mesozoic Sea Dragons: Triassic Marine Life from the Ancient Tropical Lagoon of Monte San Giorgio.* Indiana University Press, 2019. https://iupress.org/9780253040114/mesozoic -sea-dragons/.

Sciunnach, Dario. Maurizio Gaetani, and Guido Roghi. "La successione terrigena pre-Ladinica tra Lugano e Varese (Canton Ticino, Svizzera; Lombardia, Italia)." *Geologia Insubrica* (2015). https://www .academia.edu/19811699/LA_SUCCESSIONE _TERRIGENA_PRE_LADINICA_TRA _LUGANO_E_VARESE_CANTON_TICINO _SVIZZERA_LOMBARDIA_ITALIA_.

Società Italiana di Malacologia, Editors. "Fossili." S.I.M. Società Italiana di Malacologia A.P.S., June 27, 2012. https://www.societaitalianadimalacologia .it/en/publications/risorse/fossili.html.

Sommaruga, A., P. A. Hochuli, and J. Mosar. "The Middle Triassic (Anisian) Conglomerates from Capo San Martino, South of Lugano-Paradiso (Southern Alps, Switzerland)." *Geologia Insubrica* 00, no. 00 (1997): 000–000.

Stockar, Dr. Rudolf. "Facies, Depositional Environment, and Paleoecology of the Middle Triassic Cassina Beds Meride Limestone, Monte San Giorgio, Switzerland." *Swiss Journal of Geosciences* (July 2010). https://doi.org/10.1007/s00015-010-0008-2.

Strekeisen, Alex. "Ammonitico Rosso." Alex Strekeisen, 2020. http://www.alexstrekeisen.it/english/sedi /ammoniticorosso.php.

Switek, Brian. "Mosasaur Bites and Limpet Scrapes." *Wired*, Apr 11, 2012. https://www.wired.com /2012/04/mosasaur-bites-and-limpet-scrapes/.

Ticino Turismo, eds. "Monte San Giorgio." Rezzonico Editore, 2022. https://www.ticinotopten.ch /en/views/monte-san-giorgio.

Ticino.ch, eds. "Monte San Giorgio, a sea of memories." Agenzia turistica ticinese SA, 2022. https://www.ticino.ch/en/commons/details /Monte-San-Giorgio-a-sea-of-memories/73495 .html.

Tozer, E. T. "Canadian Triassic Ammonoid Faunas." *Geological Survey of Canada Bulletin* (1994). https://doi.org/10.4095/194325.

UNESCO World Heritage Centre, eds. "Nomination of Monte San Giorgio for Inclusion in the World Heritage List." UNESCO World Heritage Centre, 2003. https://whc.unesco.org.

UNESCO World Heritage Centre, eds. "Nomination of Monte San Giorgio (Italian Extension of Monte San Giorgio, Switzerland, Inscribed in 2003) for Inscription on the UNESCO World Heritage List." UNESCO World Heritage Centre, 2010. https://whc.unesco.org.

World Wildlife Federation, eds. "Shells (Deposits on the Beach)." World Wildlife Federation, n.d. https:// www.wwf.ch/fr/guide-souvenirs/coquillages -depots-sur-la-plage#guide-content.

## Chapter 11: Heteromorphs Are a Wacky Part of Evolution

Akarenga, eds. "Hokkaido's History, Culture and Nature." Akarenga, n.d. https://www.akarenga-h .jp/en/hokkaido/nature/n-01/.

Arkhipkin, Alexander I. "Getting Hooked: The Role of a U-Shaped Body Chamber in the Shell of Adult Heteromorph Ammonites." *Journal of Molluscan Studies* 80, no. 4 (November 2014): 354–364. https://doi.org/10.1093/mollus/eyu019.

Boyles, Michael J., and Alan J. Scott. "Comparison of Wave-Dominated Deltaic Deposits and Associated Sand-Rich Strand Plains, Mesaverde Group, Northwest Colorado." American Association of Petroleum Geologists, 1997. https://archives .datapages.com/data/bulletns/1982-83/data /pg/0066/0005/0550/0551c.htm.

cleanbreaks719 "Discoveries amongst the Pierre Shale of Colorado Springs." *Fossil Forum*, September 19, 2017. http://www.thefossilforum.com/index.php?/topic/77915-discoveries-amongst-the-pierre-shale-of-colorado-springs/.

Da Gama, Rui O.B.P., Brendan Lutz, Patricio Desjardins, Michelle Thompson, Iain Prince, and Irene Espejo. "Integrated Paleoenvironmental Analysis of the Niobrara Formation: Cretaceous Western Interior Seaway, Northern Colorado." *Palaeogeography, Palaeoclimatology, Palaeoecology* (November 2014). https://doi.org/10.1016/j.palaeo.2014.05.005.

David In Japan. "Fossils from Japan." *Fossil Forum*, December 21, 2022. https://www.thefossilforum.com/topic/128619-fossils-from-japan/.

Everhart, Mike. "Fossils from the Late Cretaceous Western Interior Sea." Oceans of Kansas, December 20, 2017. http://oceansofkansas.com/.

Frazier, William J., and David R. Schwimmer. "The Tejas Sequence: Tertiary—Recent." *Regional Stratigraphy of North America* (1987). https://doi.org/10.1007/978-1-4613-1795-1_9.

Fukasaw, Hiroshi. "Combine Ammonite (and Other Fossil-) Hunting and White-Water Rafting in Mukawa-cho, Hokkaido." *Education in Japan*, n.d. https://educationinjapan.wordpress.com/fielding-field-trips/combine-ammonite-and-other-fossil-hunting-and-white-water-rafting-in-mukawa-cho-hokkaido/.

Grulke, Wolfgang. "Heteromorph: The Rarest Fossil Ammonites: Nature at Its Most Bizarre." At One Communications, 2014.

Huston Joe. Heteromorph ammonite collector. Interviewed by the author, July 2022.

Kauffman, Erie G. "Paleobiogeography and Evolutionary Response Dynamic in the Cretaceous Western Interior Seaway of North America." *Semantic Scholar*, 2009. https://api.semanticscholar.org/CorpusID:221081920.

Kennedy, W. J., N. H. Landman, W. A. Cobban, and G. R. Scott. "Late Campanian (Cretaceous) Heteromorph Ammonites from the Western Interior of the United States." *Bulletin of the American Museum of Natural History* 251 (April 2000): 1–86. https://doi.org/10.1206/0003-0090(2000)251<0001:LCCHAF>2.0.CO;2.

Landman, Neil. Curator emeritus, fossil invertebrates, Division of Paleontology, Richard Gilder Graduate School of the American Museum of Natural History. Email exchange with author, January 7, 2023.

Larson, Neal L. *Ammonites and the Other Cephalopods of the Pierre Seaway: Identification Guide*. Geoscience Press, 1997.

Larson, Neal L., Jamie Brezina, Neil H. Landman, Matthew P. Garb, and Kimberly C. Handle. "Hydrocarbon Seeps: Unique Habitats That Preserved the Diversity of Fauna in the Late Cretaceous Western Interior Seaway." Geological Society of Wyoming Field Guide, 2013. https://www.academia.edu/4641897/Hydrocarbon_seeps_unique_habitats_that_preserved_the_diversity_of_fauna_in_the_Late_Cretaceous_Western_Interior_Seaway.

Lindgren, Anthony. Heteromorph ammonite collector. Email exchange with author, March 16, 2023.

Linh, Thuy. "Export of Cephalopods to Japan Show Some Positive Signal." South Pacific Fishery, n.d. https://sofis.vn/news/export-of-cephalopods-to-japan-show-some-positive-signal.

Lowery, Christopher M., R. Mark Leckie, Raquel Bryant, et al. "The Late Cretaceous Western Interior Seaway as a Model for Oxygenation Change in Epicontinental Restricted Basins." *Earth-Science Reviews*, February 1, 2018. https://doi.org/10.1016/j.earscirev.2017.12.001.

Ludvigsen, Rolf, and Graham Beard. *West Coast Fossils: A Guide to the Ancient Life of Vancouver Island*. Harbour, 1997.

Monks, Neale. "Ammonite Maturity, Pathology and Old Age." *The Cephalopod Page*, n.d. http://www.thecephalopodpage.org/ammonage.php.

Monks, Neale. "A Broad Brush History of the Cephalopoda." *The Cephalopod Page*, n.d. http://www.thecephalopodpage.org/evolution.php.

Monks, Neale. "Hooks, Paperclips and Balls of String: Understanding Heteromorph Ammonites." *Deposits Magazine*, March 16, 2017. https://depositsmag.com/2017/03/16/understanding-heteromorph-ammonites/.

Monks, Neale, and Philip Palmer. *Ammonites*. Smithsonian Institution Press, 2002.

Monks, Neale, and Jeremy R. Young. "Heteromorph Ammonites." Palaeontological Association, January 28, 1998; 1.1.1A. https://doi.org/10.26879/98001.

Monroe, James S., and Reed Wicander. *The Changing Earth: Exploring Geology and Evolution*, 7th ed. Cengage Unlimited, 2015.

Moss, Rycroft G. "Bulletin 19: The Geology of Ness and Hodgeman Counties, Kansas." University of Kansas, Lawrence, December 1932. https://www.kgs.ku.edu.

Nicholls, Elizabeth L., and Anthony P. Russell, "Paleobiogeography of the Cretaceous Western Interior Seaway of North America: The Vertebrate Evidence." *Palaeogeography, Palaeoclimatology, Palaeoecology* 79, no. 1–2 (1990). https://doi.org/10.1016/0031-0182(90)90110-S.

Palaeontologia Electronica, eds. "Heteromorph Ammonites: Animations." *Palaeontologia Electronica*, 1998–2022. https://palaeo-electronica.org/content/1-1-hetemorph-ammonites/172-1998-1/460-hetermorph-ammonites-animations#a2.

Pasch, Anne D., and Kevin C. May. "Taphonomy and Paleoenvironment of Hadrosaur (Dinosauria) from the Matanuska Formation (Turonian) in South-Central Alaska." Alaska Division of Geological and Geophysical Surveys, 1998. https://doi.org/10.14509/2335.

Prothero, Donald R. *Bringing Fossils to Life: An Introduction to Paleobiology*. Columbia University Press, 2013.

Rowe, Alison J., Neil H. Landman, J. Kirk Cochran, James D. Witts, and Matthew P. Garb. "Late Cretaceous Methane Seeps as Habitats for Newly Hatched Ammonites." *Palaios* 35, no. 3 (March 26, 2020). https://doi.org/10.2110/palo.2019.105.

Shigeta, Yasunari, and Haruyoshi Maeda. "Yezo Group Research in Sakhalin: A Historical Review." *National Science Museum Monographs* 31 (January 24, 2005). https://www.kahaku.go.jp.

Sinopaleus. "Cretaceous Ammonites of Japan." *Fossil Forum*, July 22, 2019. http://www.thefossilforum.com/index.php?/topic/96931-cretaceous-ammonites-of-japan/.

Slattery, Joshua, William A. Cobban, Kevin C. McKinney, Ashley L. Sandness, and Peter Jürgen Harries. "Early Cretaceous to Paleocene Paleogeography of the Western Interior Seaway: The Interaction of Eustasy and Tectonism." *Wyoming Geological Association 68th Annual Field Conference* 68 (June 2013). https://doi.org/10.13140/RG.2.1.4439.8801.

Stanley, Steven M., and John A. Luczaj. *Earth system history*, 4th ed. Macmillan Learning.com, 2015.

Haire, Steve. Terrestrial Treasures. Interviewed by the author, January 2023.

Takashima, Reishi, Fumihisa Kawabe, Hiroshi Nishi, Kazuyoshi Moriya, Ryoji Wani, and Hisao Ando. "Geology and Stratigraphy of Forearc Basin Sediments in Hokkaido, Japan: Cretaceous Environmental Events on the North-West Pacific Margin." *Cretaceous Research* 25, no. 3 (2004). https://doi.org/10.1016/j.cretres.2004.02.004.

U.S. National Science Foundation, eds. "Deeply Buried Sediments Tell Story of Sudden Mass Extinction." U.S. National Science Foundation, June 25, 2003. https://new.nsf.gov/news/deeply-buried-sediments-tell-story-sudden-mass.

Watanabe, Hiroto. "'Elegant' Ammonite Fossil Discovered in Hokkaido Town Identified as New Species." *The Mainichi Newspapers*, January 27, 2021. https://mainichi.jp/english/articles/20210126/p2a/00m/0sc/022000c.

Weimer, R.J., and J. S. Schlee. "Relationship of Unconformities, Tectonics, and Sea Level Changes in the Cretaceous of the Western Interior, United States, Paleotectonics and Sedimentation in the Rocky Mountain Region, United States." American Association of Petroleum Geologists, 1986. https://doi.org/10.1306/M41456C19.

Workman, Daniel. "Japan's Top 10 Exports." *World's Top Exports*, March 2023. https://www.worldstopexports.com/japans-top-10-exports/.

## Chapter 12: Ammonites Are Sacred

Adhikari, Priyanka. "Indian Tourists Make up Majority of Visitors to Muktinath." *Himalayan Times*, May 19, 2018.

Archaeology Magazine, eds. "Update on 2,600-Year-Old Celtic Grave Discovered in Germany." Archaeological Institute of America, January 26, 2017. https://www.archaeology.org/news/5230-170126-germany-celtic-tomb.

Ash, Russell. "Folklore, Myths and Legends of Britain." *Reader's Digest*, October 3, 1977.

Bath and North East Somerset, eds. "Keynsham Conservation Area Appraisal." Bath and North East Somerset, December 2016. https://www.bathnes.gov.uk/sites/default/files/sitedocuments/Planning-and-Building-Control/Planning-Policy/Placemaking-Plan/keynsham_conservation_area_appraisal.pdf.

Bolton, Aaron. "'Buffalo Stone' Could Give Insight into Blackfeet History." *Montana Public Radio*, September 10, 2019. https://www.mtpr.org/montana-news/2019-09-10/buffalo-stone-could-give-insight-into-blackfeet-history.

British Geological Survey, eds. "Lias Group: Engineering Geological Studies of Bedrock Formations." British Geological Survey, 2022. https://www.bgs.ac.uk/research/engineeringGeology/ggpp/lias_group.html.

Buckland, William. "Geology and Mineralogy Considered with Reference to Natural Theology." Ayer Company Publishers, July 1, 1980. https://doi.org/10.5962/bhl.title.36097.

Butler, Alban. *Butler's Lives of the Saints*. Loreto, 1956.

Byron Blessed. Geologist and paleobiologist, Natural Wonders Ltd. Interview with author, January 26, 2023.

Carnall, Mark. "The Ancient Mystery of St Hilda's 'Snake Stones': What Do Ammonites Really Look Like?" *The Guardian*, June 14, 2017. https://www.theguardian.com/science/2017/jun/14/the-ancient-mystery-of-st-hildas-snake-stones-what-do-ammonites-really-look-like.

Cartwright, Jane. *Feminine Sanctity and Spirituality in Medieval Wales*. University of Wales Press, 2008.

Catholic Online, eds. "St. Keyne." *Catholic Online*, n.d. https://www.catholic.org/saints/saint.php?saint_id=702.

Cho, Anjie. "Top 10 Feng Shui Crystals." *The Spruce*, December 23, 2020. https://www.thespruce.com/ammonites-used-in-feng-shui-1274355.

Costantino, Grace. "The First Described and Validly Named Dinosaur: Megalosaurus." *Biodiversity Heritage Library*, October 15, 2015. https://blog.biodiversitylibrary.org/2015/10/the-first-described-and-validly-named-dinosaur-megalosaurus.html.

Cush, Denise, Catherine Robinson, and Michael York. *Encyclopedia of Hinduism*. Routledge, 2009.

Ellin, Beth. "Great Tips for Finding Whitby Fossils." *Visit Whitby*, May 15, 2018. https://www.visitwhitby.com/blog/great-tips-finding-whitby-fossils/.

Ford, David Nash. "St. Keyne Wyry." Nash Ford, 2001. https://www.earlybritishkingdoms.com/bios/keynewbg.html.

fulltimeexplorer. "The 21 Best Nepal Souvenirs & Where to Buy Them." *Fulltimeexplorer*, July 3, 2023. https://fulltimeexplorer.com/nepal-souvenirs-kathmandu/.

Gadacz, René R. "Medicine Bundles." *Canadian Encyclopedia*, 2006. Last updated November 15, 2019. https://www.thecanadianencyclopedia.ca/en/article/medicine-bundles.

Grimes, John A. *A Concise Dictionary of Indian Philosophy: Sanskrit Terms Defined in English.* State University of New York Press, 1996.

Grinnell, George Bird. *Blackfeet Indian Stories.* Scribner's, 1913.

Jakimowicz-Shah, Marta. *Metamorphoses of Indian Gods.* Seagull, 1988.

Johnston, David. *Discovering Roman Britain, Book No. 272.* Shire, 1983.

Jones, Constance, and James D. Ryan. *Encyclopedia of Hinduism.* Infobase, 2006.

Kalama, Wailana. "A Legend of Snakes and Stones." Tula Foundation and Hakai Institute, July 26, 2018. https://www.hakaimagazine.com/article-short/a-legend-of-snakes-and-stones/.

Kiefer, James E. "Hilda of Whitby, Abbess and Peacemaker." Society of Archbishop Justus, 1999. http://justus.anglican.org/resources/bio/285.html.

Knowlton, Troy. Lifelong ammonite hunter, member of the Blackfoot Confederacy Piikani nation. Telephone interview with author, April 3, 2022.

Korite.com, eds. "The Blackfoot 'Buffalo Stone' and KORITE Ammolite." *Korite*, June 21, 2021. https://korite.com/blogs/news/the-blackfoot-buffalo-stone-and-korite-ammolite-the-legend-heritage.

Krishna, Nanditha. *Hinduism and Nature.* Penguin Random House India, 2017.

Lochtefeld, James G. *The Illustrated Encyclopedia of Hinduism.* Rosen, 2002.

Lomax, Dean R. *Fossils of the Whitby Coast: A Photographic Guide.* Siri Scientific Press, 2011.

Lotzof, Kerry. "What on Earth? Snakestones: The Myth, Magic and Science of Ammonites." Trustees of The Natural History Museum, London, Undated. https://www.nhm.ac.uk/discover/snakestones-ammonites-myth-magic-science.html.

Messerschmidt, Don. "In the Presence of the Gods: Shaligram Pilgrimage in the Nepal Himalaya." Himal Southasian, July 6, 2021. https://www.himalmag.com/in-the-presence-of-the-gods-shaligram-pilgrimage-review-2021/.

Monier-Williams, Monier. *A Sanskrit-English Dictionary: Etymologically and Philologically Arranged with Special Reference to Cognate Indo-European Languages.* Motilal Banarsidass, 1899

Morton, Nicol, and Stephen Hesselbo. "International Subcommission on Jurassic Stratigraphy. Newsletter 35/1." *Jurassic Earth*, December 2008. http://jurassic.earth.ox.ac.uk/.

Motologs. "Muktinath Travel Guide—India to Nepal Road Trip." *Motologs.com*, May 16, 2023. https://motologs.com/muktinath-travel-guide-india-nepal-road-trip/.

Naish, Darren, and David M. Martill. "Dinosaurs of Great Britain and the Role of the Geological Society of London in Their Discovery: Basal Dinosauria and Saurischia." *Journal of the Geological Society* 164, no. 3 (May 2007). https://doi.org/10.1144/0016-76492006-032.

National Trust, eds. "Fossil Hunting on the Yorkshire Coast." *National Trust*, n.d. https://www.nationaltrust.org.uk/features/fossil-hunting-on-the-yorkshire-coast.

NepaliSansar, ed. "Nepal's Top Pilgrimage and Holy Sites: The Abode of Spirituality." *Nepali Sansar*, March 30, 2019. https://www.nepalisansar.com/tourism/nepals-top-pilgrimage-sites-the-abode-of-spirituality/.

Northern Saints, eds. "Trails." Visit County Durham, n.d. https://www.thisisdurham.com/northernsaints/trails.

Order of the Holy Paraclete, The, eds. "St. Hilda of Whitby." Order of the Holy Paraclete, n.d. https://www.ohpwhitby.org.uk/the-priory/st-hilda-of-whitby/.

painshill. "Buffalo Calling Stone (Iniskim) & Blackfeet Legend." *Arrowheads*, December 4, 2013. https://forums.arrowheads.com/forum/information-center-gc33/lithic-artifacts-technology-materials-gc71/gamestones-charmstones-effigy-items-gc79/106701-buffalo-calling-stone-iniskim-blackfeet-legend.

Paleontica.org, eds. "North Yorkshire coast." Stichting Paleontica, n.d. https://www.paleontica.org/sites/fossil_site.php?plaats=85&language=en.

Parish Church of St. Wilfrid, Bognor, eds. "St. Hilda." *St. Wilfrid's Parish Magazine*, September 2020. https://static1.squarespace.com/static/5eb00334e6945a698f17aff0/t/5f8613578634c46625aabe11/1602622397917/magazine202009.pdf.

Paul Taylor. Scientific associate and emeritus merit researcher, editor-in-chief, Palaeontological Association, Natural History Museum. Email dialogue with author, January 24, 2023.

Rising Nepal, The, eds. "The Import Scenario." *Rising Nepal*, April 21, 2023. https://www.risingnepaldaily.com/news/25695.

Robinson, Stephen. *Somerset Place Names*. Dovecote Press, 1992.

Rudwick, Martin J. S. *Scenes from Deep Time: Early Pictorial Representations of the Prehistoric World*. University of Chicago Press, 1992.

Rushby, Kevin. "Walking with Dinosaurs: Fossil Hunting on the North Yorkshire Coast." *The Guardian*, March 11, 2017. https://www.theguardian.com/travel/2017/mar/11/north-yorkshire-coast-beach-dinosaur-fossil-hunting-saltwick-bay.

Shastri, J. L. *The Shiva Purana*. Motilal Banarsidass, 2014.

Shoreline Cottages, eds. "Whitby History & Mythical Tales: Whitby Jet." *Shoreline Cottages*, 2023. https://shoreline-cottages.com/whitby-life-blog/whitby-jet/.

Spencer, Ray. *A Guide to the Saints of Wales and the West Country*. Llanerch Press, 1991.

Stone Science. Facebook Post. Facebook, April 17, 2021. https://www.facebook.com/StoneScienceAnglesey/photos/weve-been-looking-at-myths-and-legends-concerning-stones-and-crystals-what-about/2957317461255635/.

Taylor, Paul. Email dialogue with author, January 24, 2023.

Taylor, Paul D. "Fossil Folklore: Ammonites." *Deposits Mag*, September 27, 2016. https://depositsmag.com/2016/09/27/fossil-folklore-ammonites/.

Telus Storyhive. "Ammolite: Gem of the West." *YouTube.com*, Undated. https://www.youtube.com/watch?v=jr9kV90oS_w&t=14s.

Tutcher, J. W. *Yorkshire Type Ammonites*. William Wesley and Son, 1912. http://www.archive.org/stream/yorkshiretypeamm01buckuoft\.

UK Fossils, eds. "UK Fossil Collecting, Yorkshire, Robin Hoods Bay." *UK Fossils*, 2022. https://ukfossils.co.uk/2007/03/18/robin-hoods-bay/.

UK Fossils, eds. "UK Fossil Collecting, Yorkshire, Whitby." *UK Fossils*, 2022. https://ukfossils.co.uk/2007/03/18/whitby/.

Vijayalakshmy, R. "An Introduction to Religion and Philosophy: Tévarám and Tivviyappirapantam." International Institute of Tamil Studies. https://www.exoticindiaart.com/book/details/introduction-to-religion-and-philosophy-tevaram-and-tiviyappirapantam-old-and-rare-book-uan031/.

Walters, Holly. *Shaligram Pilgrimage in the Nepal Himalayas*. Amsterdam University Press, 2020.

Wearn, Rebecca. "Tourism Is the Lifeblood of the Nepalese Economy." *BBC News*, April 18, 2016. https://www.bbc.com/news/business-35695602.

*Whitby Guide, The*, eds. "Whitby Fossils: How and Where to Find Fossils in and Around Whitby."

*The Whitby Guide*, January 25, 2022. https://www.thewhitbyguide.co.uk/whitby-fossils/.

Whitby Museum, eds. "Geology and Fossils." Whitby Museum, 2022. https://whitbymuseum.org.uk/collection/geology-and-fossils/.

Whitby-Yorkshire.County UK., eds. Folklore and Legends: St Hilda." *Whitby-Yorkshire.co.UK*, 1996–2022. http://www.whitby-yorkshire.co.uk/folklore/folklore.htm.

Wikisource. 2011. "Page: Folklore: A Quarterly Review. Volume 16, 1905.djvu/384." Wikisource.com. Last updated August 19, 2018. https://en.wikisource.org/wiki/Page:Folk-lore_-_A_Quarterly_Review._Volume_16,_1905.djvu/384.

World Bank, eds. "Harnessing Tourism to Enhance the Value of Biodiversity and Promote Conservation in Nepal." World Bank, June 3,2022. https://www.worldbank.org/en/news/feature/2022/06/03/harnessing-tourism-to-enhance-the-value-of-biodiversity-and-promote-conservation-in-nepal.

Zurick, David, and Julsun Pacheco. *Illustrated Atlas of the Himalaya*. University Press of Kentucky, August 4, 2006.

## Chapter 13: They Arrived after One Extinction and Expired during Another

AMNH, eds. "Antarctica's Wealth of Ammonites." American Museum of Natural History (AMNH), June 22, 2018. https://www.amnh.org/explore/news-blogs/news-posts/antarctica-ammonites.

Batt, Richard J. "Ammonite Shell Morphotype Distributions in the Western Interior Greenhorn Sea and Some Paleoecological Implications." *Palaios* 4, no. 1 (1989): 32–42. https://doi.org/10.2307/3514731.

*Encyclopaedia Britannica*, eds. "Seymour Island." *Encyclopedia Britannica*, n.d. https://www.britannica.com/place/Seymour-Island-Weddell-Sea.

Byron Blessed. Geologist and paleobiologist, Natural Wonders Ltd. Interview with author, January 26, 2023.

Choi, Charles Q. "Digging Up Valuable Fossils in Suburban New Jersey." *Scientific American*, 2022. https://www.scientificamerican.com/article/digging-up-valuable-fossils-in-nj/.

DePalma, Robert A., Jan Smit, David A. Burnham, and Walter Alvarez. "A Seismically Induced Onshore Surge Deposit at the KPg Boundary, North Dakota." *PNAS*, April 1, 2019. https://doi.org/10.1073/pnas.1817407116.

frankh8147. "The Pinna Layer of Central New Jersey (USA)." *Fossil Forum*, December 20, 2020. http://www.thefossilforum.com/index.php?/topic/111523-the-pinna-layer-of-central-new-jersey-usa.

Greshko, Michael, and National Geographic Staff. "What Are Mass Extinctions, and What Causes Them?" *National Geographic*, September 26, 2019. www.nationalgeographic.com/science/prehistoric-world/mass-extinction.

Klug, Christian. Email exchange with author, February 14, 2023.

Kowinsky, Jayson. "Hell Creek Dinosaur Fossil Hunt." *Fossilguy.com*. 2022. https://www.fossilguy.com/trips/hell-creek-2020/index.htm.

Kowinsky, Jayson. "The Hell Creek Fossil Hunting Trip." *Fossilguy.com*, 2022. https://www.fossilguy.com/trips/hell-creek-2017/index.htm.

Landman, N., S. Goolaerts, J. Jagt, E. Jagt-Yazykova, and M. Machalski. "Ammonites on the Brink of Extinction: Diversity, Abundance, and Ecology of the Order Ammonoidea at the Cretaceous/Paleogene (K/Pg) Boundary." Ammonoid Paleobiology: From macroevolution to paleogeography. Springer, 2015. Topics in Geobiology, vol 44. https://doi.org/10.1007/978-94-017-9633-0_19.

Landman, Neil H., Ralph O. Johnson, Matthew P. Garb, Lucy E. Edwards, and Frank Thomas Kyte. "Cephalopods from the Cretaceous/Tertiary

Boundary Interval on the Atlantic Coastal Plain, with a Description of the Highest Ammonite Zones in North America. Part 3, Manasquan River Basin, Monmouth County, New Jersey." *Bulletin of the American Museum of Natural History*, no. 303 (2007). http://hdl.handle.net/2246/5850.

Landman, Neil H., Stijn Goolaerts, John W. M. Jagt, Elena A. Jagt-Yazykova, Marcin Machalski, and Margaret M. Yacobucci. "Ammonite Extinction and Nautilid Survival at the End of the Cretaceous." *Geology* 42, no. 8 (2014): 707–710. https://doi.org/10.1130/G35776.1.

Macellari, Carlos E. "Late Campanian-Maastrichtian Ammonite Fauna from Seymour Island (Antarctic Peninsula)." *Journal of Paleontology* 60, no. S18 (March 18, 1986): 1–55. https://doi.org/10.1017/S0022336000060765.

Machalski, Marcin, and Claus Heinberg. "Evidence for Ammonite Survival into the Danian (Paleogene) from the Cerithium Limestone at Stevns Klint, Denmark." *Bulletin of the Geological Society of Denmark*, December 2005. https://www.researchgate.net/publication/261359356.

Landman, Neil. Curator emeritus, fossil invertebrates, Division of Paleontology, Richard Gilder Graduate School of the American Museum of Natural History in New York City. Email exchange with author, September 2021.

Premovic, Pavle, Bratislav Ž. Todorović, and Mirjana Pavlovic. "Cretaceous—Paleogene Boundary Fish Clay at Højerup (Stevns Klint, Denmark): Trace Metals in Kerogen." *Bulletin de la Société Géologique de France* 178, no. 5 (September 2007): 411–421. https://doi.org/10.2113/gssgfbull.178.5.411. https://www.researchgate.net/publication/270447455.

Sci.News. News Staff/Source. "66-Million-Year-Old Fossil Site Preserves Animals Killed within Minutes of Chicxulub Impact." *Science News*, April 3, 2019. http://www.sci-news.com/paleontology/tanis-fossil-site-07055.html.

Wikipedia. "Extinction Event." *Wikipedia.com*, updated October 19, 2022. https://en.wikipedia.org/wiki/Extinction_event.

Wikipedia. "Stevnsfortet." *Wikipedia.com*, updated January 2, 2022. https://da.wikipedia.org/wiki/Stevnsfortet.

Wikipedia. "Stevns Klint." *Wikipedia.com*, updated August 24, 2022. https://en.wikipedia.org/wiki/Stevns_Klint#UNESCO_listing.

Wikipedia. "Tanis (fossil site) 2019." *Wikipedia.com*. updated October 19, 2022. https://en.wikipedia.org/wiki/Tanis_(fossil_site).

Zinsmeister, William J., Rodney M. Feldmann, Michael O. Woodburne, and David H. Elliot. "Latest Cretaceous/Earliest Tertiary Transition on Seymour Island, Antarctica." *Journal of Paleontology* 63, no. 6., 1989: 731–738. http://www.jstor.org/stable/1305641.

## Chapter 14: What You End Up Collecting

American Museum of Natural History. "Velociraptor Had Feathers." *ScienceDaily*, September 20, 2007. https://www.sciencedaily.com/releases/2007/09/070920145402.htm.

Bylund, Kevin. "Ammonoid Family Reunions." *Ammonoidea*, February 11, 2021. http://ammonoidea.blogspot.com/2017/11/ammonoid-family-reunions.html.

Greshko, Michael. "Spectacular New Fossil Bonanza Captures Explosion of Early Life." *National Geographic*, March 21, 2019. https://www.nationalgeographic.com/science/article/treasure-trove-of-spectacular-fossils-found-in-china.

Jattiot, Romain, Hugo Bucher, Arnaud Brayard, Claude Monnet, James F. Jenks, and Michael Hautmann. "Revision of the Genus Anasibirites Mojsisovics (Ammonoidea): An Iconic and Cosmopolitan Taxon of the Late Smithian (Early Triassic) Extinction." Pap Palaeontology 2 (November 17, 2015): 155–188. https://doi.org/10.1002/spp2.1036.

Jurassic Coast World Heritage Site Partnership, eds. "What Is the Jurassic Coast?" Jurassic Coast World Heritage Site Partnership, 2022. https://jurassiccoast.org/.

Kerner, Andreas. President of International Fossil Co. Email dialogue with author, November 5, 2022.

Larson, Dennis. "How to Check if That Fossil Is Legal." Fossil Hunters, September 5, 2022. https://www.fossilhunters.xyz/fossil-industry/appendix-how-to-check-if-that-fossil-is-legal.html.

Liston, J. J., and Hai-Lu You. "Chinese Fossil Protection Law and the Illegal Export of Vertebrate Fossils from China." BioOne, March 2015.

Natural History Museum (NHM) of Los Angeles County. "Meet Your Museum: Dr. Austin Hendy, Geology & Paleontology." *YouTube.com*, January 28, 2021. https://www.youtube.com/watch?v=X0sOiaZQM2o&t=338s.

Sample, Ian. "'Mind-Blowing' Haul of Fossils over 500m Years Old Unearthed in China." *The Guardian*, March 21, 2019. https://www.theguardian.com/science/2019/mar/21/mindblowing-haul-of-fossils-over-500m-years-old-unearthed-in-china.

Tarpy, Cliff. "Liaoning Province—China's Extraordinary Fossil Site." *National Geographic*, No date. https://www.nationalgeographic.com/science/article/china-fossils.

UNESCO, eds. "Leye Fengshan UNESCO Global Geopark (China)." *UNESCO*, 2021. https://en.unesco.org/global-geoparks/leye-fengshan.

Wikipedia. "Early Triassic." *Wikipedia.com*, last updated August 30, 2022. https://en.wikipedia.org/wiki/Early_Triassic#Smithian-Spathian_boundary_extinction.

# Index

Page numbers in *italics* refer to illustrations.

*Calliphylloceras nilssoni*, 172
*Caloceras* (*Psiloceras*), 205
*Calycoceras*, 129; *tarrantense*, 3
Cambrian period, 10
Camden, William, 57, 205
camerae, 4, 23, 125–126
Canada, 196
*Cancelloceras*, 126
Cannonball Formation, 104
cannonballs (ammonite geodes), *208*
canyons, 168
Capo San Vigilio, Italy, 172–174
*Cardioceras cordatum*, 143
Carew, Richard, 200
carrier snail (*Xenophora*), *186*
Cascade Mountains, 72
Catalani, John, 48
Catlin, George, *198*
Catskill Delta, 17
Cenozoic Era, 222–223
cephalopods: ammonites sutures and, 68; ammonites varieties of, *20*; *Ammonoidea* as subclass of, 3; ammonoids as, 3, 114; Aveyron celebrating, 143; bread recipe, 142; during Cambrian explosion, 10; Chicxulub asteroid killing off, 215; coleoid, *122*; collectors interest in, xiv–xv; eating apparatus of, 114; fossils of, 35; Late Cretaceous, 187–189; life span of, 137; nodes of, 67; PPF intermingled with, 91; reproductive capacity of, 138; along Rio Napo, 92; shells of, 58, 152–153; subclass of, 3, 114; *wanisugna*, 199
*Ceratites*, 27, *125*; *enodis*, 147; *evolutus*, 145; *munsteri*, 147; *nodosus*, 145; *pulcher* zone, 147; *semipartitus*, 145
Ceratitida, 27, 51
chambered nautilus (*Nautilus pompilius*), *115*
Chang Qu, 239
Charles M. Russell National Wildlife Refuge, *224*
Charmouth Heritage Coast Centre, 63
*Cheiloceras amblylobum*, 17
*Cheloniceras*, 188
Chengjiang biota, 95
Cherns, Lesley, 116
Cherry Valley Limestone Lagerstätte, 13, 15–16
Chesil Beach, West Bay, *64*
Chicxulub asteroid, 219–221; extinction patterns after, 217; ocean acidification after, 215–216; species eliminated by, 152; at Yucatan Peninsula, *153*, 214
Children of Ammon, 36
China, 239–246
Chinle Formation, 80
chitinous beak, 124
*Choffaticeras segne*, *91*
Choteč Event, 23
Church Cliff, 61–62
Civic Museum of Natural History, 177
*Cleoniceras*, *6*, *101*, *124*; lemon calcite conservation, *83*; opalescent nacre of, 105, *105*; undulating pyrite suture design of, 98
Cleopatra, 36
*Cleviceras. See Eleganticeras*
*Clionites*, 70
*Clionitites*, *73*

*Clioscaphites*, 187
Closs, Gerhard Ludwig, 113
*Clymenia*, 25, 40, 67
Clymeniida, 24, 45, 48–50
*Colchidites breistrofferi*, *97*
coleoid cephalopods, *122*
*Coleoidea*, 114
*Collignoniceras woollgari*, 226
*Colombiceras*, 188
common squid (*Loligo vulgaris*), *115*
*Compsognathus*, 106
conchs, 3, 10; geometry, 75, *178*; goniatites in front of external, 23; marine animals with, 12; trace elements in, 102
concretion, 79–80, *80*, 96, *96*
consumerism, in China, 246
Coober Pedy, Australia, 87
Cook, James, 204
Cooper, Ben, 235
coprolite, 151, *151–152*
Coquina, 91
corkscrew helixes, 12
*Corongoceras*, 194
*Coroniceras*, 141, *141*; *Dalle*, 142; *multicostatum*, 140
Coropuna peak, 128, *129*
Cotahuasi Canyon, in Peru, 128–131, *129*
Cotahuasi River, 128, *129*
Cotahuasi Subbasin Landscape Reserve, 130–131
Cottonwood Creek, Northern California, *230*
Cottrell, John, 13
*Cranocephalites costidensus*, 222
*Craspedodiscus*, 109
Cretaceous conifers, 90
Cretaceous-Paleogene boundaries, 215–216, 220, 226
Cretaceous-Paleogene (K-Pg) extinction event, 214, *215*, 216
crinoids, 44, 85–86, *94*, 151
*Crioceratites*, 161, 180; *duvalii*, *37*; *loryi*, *34*, *161*
"Crocodile in a Fossil State," 62
crystalline structure, 99–100
Cuiping village, *241*
Cuthbert (saint), 204
cuttlefish (*Sepia officinalis*), *115*, 121–122
Cuvier, Georges, 57, 205
*Cymatoceratidae*, *4*

*Dactylioceras*, *20*, 37, 65, *84*, 87; as ammonite pebble, *2*, *208*; *commune*, *201*; in Lias Group, 206; in medieval Europe, 202; *semicelatum*, 95
*Dactylioceratidae*, 86, 154
*Dakotaraptor*, 225
Daleje Transgression, 45
Dalle, The, 140–142
Darwin, Charles, 27
De Baets, Kenneth, 137–139
decay, of organic matter, *83*, 84–85
deep water squid (*Austrorossia mastigophora*), *115*
Denmark, 217–219, *218*
Denver Mineral, Fossil, Gem and Jewelry Show, 238
DePalma, Robert, 214
*Desmoceratidae*, 217, 220
Devonian limestone layer, 17

marine life, xv, 2–3, 11–12
"Marmion, a Tale of Flodden Field" (poem), 204
mass mortality events, 15, 44–45, 140
Mayo Salt Mine, 54
Mayr, Ernst, 75
Meacher, Mike, 16
*Meekoceras*, 242; *gracilitatis*, 243, *243*; Zone, 243–244
*Meekocerataceae*, 52
*Megalosaurus*, 205
*Mercaticeras*, 86
Meride limestone, 170–172
Mesozoic Era, 216, 222
Mesozoic marine reptiles, 62
*Metalegoceras sundaicum*, *150*
metaphysical powers, 191, 249
*Metatissotia*, *37*
microconch (male), 137
*Microderoceras birchi*, *134*
microorganisms, 152
*Microtropites*, 74
Mid-America Paleontology Society (MAPS), 238
Middle Cambrian Burgess Shale, 95
Middle Kingdom, 196
*Mierotropites*, 73
*Mimagoniatitidae*, 138
*Mimosphinctidae*, 137–138
minerals, 83, *83*, 97, 144
mining, for fossils, 118
Ministry of Land and Resources, 245
mollusks, 2, 58, 81
Monarrez, Pedro, 48
Montana, Bearpaw Shale in, 123
Monte San Giorgio, 169–172
Morawica Mine, 38
Morcote, Switzerland, *169*
*Moremanoceras straini*, *214*
Morocco, 41, *43*; ammonoids record in, 42; Devonian trilobite
    beds in, 95; fossil prospecting in, 47; Little Atlas Mountains
    in, 46; Tafilalt Plateau in, 42–44, *45*, 46
*Mortoniceras*, 93, *93*
*Mosasaurs*, 165–167, *166*
mountain ranges, 42, 168
Mount Shasta, 72
Mount Vesuvius (79 CE), 168
mouthpart variations, of ammonoids, 124
*Muensteroceras*, 155
*Muessenbiaergia sublaevis*, 45
Mughal Empire, 54
Mukawa River, 186
Münchner Mineralientage, 238
Munich, Germany, 238
Murray, John, 198
muscular hydrostats, 114

NaCl. *See* sodium chloride
*Nammalites*, 52, 248
Napo River, in Peru, *92*
Natural History Museum in London, 62
*Naturalis Historiae* (Pliny the Elder), 37

nautiloid, delineated central siphuncle of, *4*
*Nautiloidea*, 44
nautilus, tentacles of, 122–123
*Nautilus pompilius*, *115*, *117. See also* chambered nautilus
Navajo Nations, 199
*Nebrodites*, 158, *159*
necrosis, 93
*Neoicoceratidae*, 52
*Neolobites*, 129
*Neophlycticeras*, 93
Nepal, Federal Democratic Republic of, 195
*Nevadites*, 170
New Guinea, Upper Sepik River, 195
*Nipponites*, 180, 186
Niti Limestone, 243
*Normannites*, 136
*North American Indian Portfolio* (Catlin), *198*
North American landmasses, 181–182
North American Tectonic Plate, 71
North Yorkshire, England, *207*
*Nostoceras*, 184, *185*
Nubian Salvage Campaign, 41

obrution deposits, 94–95
ocean acidification, 215–216, 222–223
octopus (*Eledone moschata*), *115*, 121
oil-shale matrix, 85
Omnibus Land Preservations Act, 239
*Omomyidae*, 222
onychites, *122*
opalescent ammonites, *105*, 105–106
*Ophiceras*, *164*; *greenlandicum*, 125
Oppeliidae, 164
Ordovician Period, 213
Ordovician-Silurian extinction, 11
organic matter, decay of, 84–85
organisms, fossil preservation of, 100
*On the Origin of Species* (Darwin), 27
orogeny, mountain ranges from, 42
*Orthaspidoceras*, 165
Orthoceras, 7
*Orthoceratidae*, 46
*Orthocerida*, *12*, *43*, *44*
*Otoceratoidea*, 125
*Ovaticeras*, 207
*Owenites*, 54, 244, 247
oxidation, 81
oxidoreduction, 82
*Oxybeloceras meekanum*, *178*
oxygen, ocean's lack of, 44, 46–48, 79, 85–86, 94
*Oxytropidoceras*, 129

*Pachydiscus*, 166, 220
*Pachypleurosaurus, Monte San Giorgio* (journal), 171
Pacific Southwest (PSW) Research Station, 70
Pakistan: Ceratitida found in, 51; fossil prospecting in, 53;
    Jhelum, 54–55; Rohtas Fort in, *53*; Salt Range in, 50
PaleoBOND™ fossil glue, 102
paleontology: acquisitional, 234–235; vocational, 231–233